FRIENDS AND PARTNERS

FRIENDS AND PARTNERS

The Legacy of Franklin D.
Roosevelt and Basil O'Connor
in the History of Polio

DAVID W. ROSE

Series Editor
EDWARD R. B. McCABE
March of Dimes Foundation

Amsterdam • Boston • Heidelberg • London
New York • Oxford • Paris • San Diego
San Francisco • Singapore • Sydney • Tokyo
Academic Press is an imprint of Elsevier

Academic Press is an imprint of Elsevier
125 London Wall, London EC2Y 5AS, UK
525 B Street, Suite 1800, San Diego, CA 92101-4495, USA
50 Hampshire Street, 5th Floor, Cambridge, MA 02139, USA
The Boulevard, Langford Lane, Kidlington, Oxford OX5 1GB, UK

Notices

Knowledge and best practice in this field are constantly changing. As new research and
experience broaden our understanding, changes in research methods, professional practices,
or medical treatment may become necessary.

Practitioners and researchers must always rely on their own experience and knowledge in
evaluating and using any information, methods, compounds, or experiments described
herein. In using such information or methods they should be mindful of their own safety
and the safety of others, including parties for whom they have a professional responsibility.

To the fullest extent of the law, neither the Publisher nor the authors, contributors, or
editors, assume any liability for any injury and/or damage to persons or property as a
matter of products liability, negligence or otherwise, or from any use or operation of any
methods, products, instructions, or ideas contained in the material herein.

British Library Cataloguing-in-Publication Data
A catalogue record for this book is available from the British Library

Library of Congress Cataloging-in-Publication Data
A catalog record for this book is available from the Library of Congress

ISBN: 978-0-12-803597-9

Typeset by TNQ Books and Journals
www.tnq.co.in

DEDICATION

In Memory of Charles L. Massey (1922–2015)
President of the March of Dimes (1978–1989)

CONTENTS

FOREWORD

Most Americans today know little about polio—and that's surely what Basil O'Connor would have wanted. The disease he fought so hard to conquer is now a relic of history, banished to isolated corners of the globe, on the brink of eradication. It was not an easy fight. Those who lived through the polio years well remember the fear it generated and the suffering it caused. The battle had an iconic symbol in Franklin Roosevelt, a battalion of brilliant scientists led by Albert Sabin and Jonas Salk, and an army of dedicated March of Dimes volunteers. Largely ignored, however, is the field general who molded this talent and energy into a stunning national crusade. The full story of Basil O'Connor's central role in defeating polio, deftly told here by David Rose, is long overdue.

Polio is an intestinal infection that spreads from person to person through oral–fecal contact: unwashed hands, contaminated food or water, or shared objects. The agent is a virus that enters the body through the mouth, travels down the digestive tract, and is excreted in the bowels. Usually the infection it produces is slight—a headache, slight fever, nausea—though the victim remains a carrier of the disease. In rare instance the virus enters the bloodstream, invading the brain and central nervous system and destroying the nerve cells that stimulate the muscle fibers to contract. Death most often occurs when the breathing muscles controlled by the brain stem (or bulb) are involved, a condition known as bulbar polio.

Once called infantile paralysis, polio did not appear in epidemic form in the United States until 1916, claiming 6,000 lives, many of them in New York City. Why it came when it did, and why the mass outbreaks occurred almost exclusively in the United States and other developed nations, remains a mystery. Polio in *epidemic* form is both a modern phenomenon and an affliction of the West.

Some see it as a disease of cleanliness. As Western nations industrialized, they became more health conscious. Clean water, better sanitation, and stricter personal hygiene became hallmarks of modern life. What Americans could not foresee, however, was that their antiseptic revolution brought risks as well as rewards. As these reforms advanced, people were less likely to come into contact with poliovirus early in life, when the infection is milder and maternal antibodies offer temporary protection. The end result would be a vastly expanding pool of victims—the most famous, of course, being Franklin D. Roosevelt (FDR).

When Roosevelt was struck down by polio in 1921, the public response bordered on disbelief. The disease was just taking hold in the United States and the victims were mainly children, as the New York City epidemic of 1916 had so cruelly shown. Yet here was FDR—39 years old, robust and athletic—at the mercy of a children's disease, paralyzed from the waist down.

When I was writing *Polio: An American Story*, my editor asked me to explain how FDR contracted polio. I flippantly (if correctly) replied: "He was unlucky." That wouldn't do, of course. I had to provide some background to explain Roosevelt's ill fortune.

It was well known that FDR had come of age on an isolated estate in the Hudson Valley. He had been sheltered by a small army of nannies and tutors, interacting with adults and avoiding the common childhood illnesses until his arrival at boarding school. "From that point forward," I wrote, "his medical history resembled an encyclopedia of contagious diseases. The list included typhoid fever, swollen sinuses, infected tonsils, and endless sore throats, some of which forced him to bed for weeks." During the influenza pandemic of 1918–19, Roosevelt contracted a case of double pneumonia that almost took his life. His early years, it appeared, had left him "immunologically unprepared" for the world beyond Hyde Park.

There was more. The summer of 1921 had been especially hard on him—intense, humiliating, and pressure packed. Already drained from a losing campaign for vice president the previous fall, FDR found himself in the middle of a national scandal regarding his role, as assistant secretary, in allegedly using undercover agents to entrap young sailors. It was a bogus charge, politically motivated, but it forced him to testify before Congress in the brutal Washington heat. Exhausted, he set out for the family's retreat at Campobello, in Canada's Bay of Fundy, stopping along the way to attend a massive Boy Scout jamboree at New York's Bear Mountain Park. It was here, in all likelihood, that poliovirus entered his body. A poignant photo, one of the last taken of Roosevelt walking unassisted, shows him marching in the scout parade.

Upon reaching Campobello, FDR lost himself in a whirl of activity—sailing, swimming, hiking, and drinking into the night. Stress and heavy exercise can lower one's immunity. Following a relay race with his children and a swim in the frigid Bay of Fundy, FDR spent the afternoon in a wet bathing suit answering his mail. He woke up the next morning with a fever and a limp leg. Within days, he'd lost all feeling below the waist. He was paralyzed.

It would take Roosevelt 7 years to return to public life. In between, he tried every conceivable cure for paralysis, to no avail. He even purchased a broken down resort in Warm Springs, Georgia, where the soothing mineral waters held the promise of recovery, and where the idea of helping other polio survivors took shape. Warm Springs would remain an integral part of Roosevelt's life—a place where he could be himself, without having to hide his paralysis or worry about the intrusions of the press.

It was during this hiatus that Roosevelt met O'Connor. Their paths had crossed briefly in FDR's prepolio days, introduced by O'Connor's brother, a New York City politician. A more fateful meeting occurred in the lobby of the Broadway building where O'Connor worked as an attorney and Roosevelt, a lawyer himself, kept an office. Walking painfully on the arm of his chauffeur, Roosevelt slipped and fell on the polished marble floor. O'Connor was among those who rushed over to help him up.

An unlikely union evolved. The two men had little in common. O'Connor came from a Catholic working-class background. A self-made man, having worked his way through Dartmouth and Harvard Law School, he lacked Roosevelt's pedigree and easygoing charm. FDR was looking to start a law firm based on his name and connections, with someone else handling the mundane day-to-day chores; O'Connor, a notorious workaholic, seemed the ideal choice. "If Roosevelt was the young prince," said one observer, "O'Connor was the perfect vassal."

David Rose tells the full, inspiring story of this remarkable partnership. Had FDR not returned to politics, he likely would have spent his days focusing on the rehabilitation of polio survivors and the search for a cure. Instead, he chose the substitute he trusted the most. Standing next to FDR on the day he announced the formation of the National Foundation for Infantile Paralysis in 1938 was Basil O'Connor, the handpicked director, who vowed to end polio whatever the cost. "I am," said O'Connor, "very confident of our future."

It proved a brilliant selection. Stubborn and single-minded, with great organizational skills, O'Connor revolutionized the manner in which philanthropies raised money, recruited volunteers, and advertised the cause. It was the National Foundation—better known to Americans as the March of Dimes—that invented the poster child and the mothers' marches used so successfully to this day by other charities. It was the March of Dimes that first used celebrities—from Eddie Cantor to Helen Hayes to Grace Kelly to Elvis Presley—for attention-grabbing public events. And it was the March of Dimes that turned fund-raising on its head by seeking small gifts from the many rather than large gifts from the few. No one, even in the depths of the Great Depression, was too poor to give a dime to help a child walk again.

Above all, it was the March of Dimes that created a radical new model for "giving" in the United States, the concept of philanthropy as consumerism, with donors promised the ultimate personal reward: protection against the disease itself.

Polio reached its peak in America in the late 1940s and the early 1950s, with as many as 50,000 cases a year. It became, in essence, the crack in the middle-class window of an increasingly suburban, prosperous, baby-centered post–World War II culture. Polio hit without warning each summer like a plague. Newspapers kept daily box scores of those admitted to hospital polio wards. Beaches, swimming pools, bowling alleys, and movie theaters were closed. Rumors abounded that one could get polio from an unguarded sneeze, handling paper money, or talking on the telephone. "We got to the point that no one could comprehend," a pediatrician recalled, "when people would not even shake hands."

Actually, polio was never quite the raging epidemic portrayed in the press. Ten times as many children would be killed in accidents in these years, and three times as many would die of cancer. The national dread of polio had two main causes. One was the intensely visual nature of the disease. A person could walk into restaurant without knowing who might have heart disease or leukemia, but no one could miss the telltale signs of polio: crutches, leg braces, and wheel chairs. The other was the strategy of the National Foundation to hype the power of polio to strike down anyone, anywhere, at any time. There was but one way for parents to protect their children—and that was to support the March of Dimes.

Basil O'Connor knew little about science, much less the dynamics of polio. What he did know, however, was that progress on this front had been painfully slow. With abundant resources at his disposal, O'Connor convinced Tom Rivers, a pioneering virologist, to form a committee on scientific research within the Foundation. Then he hired a superb administrator named Harry Weaver to oversee the agenda. Rivers and Weaver agreed that prevention was the best path to follow—and that meant a vaccine.

Three basic questions had to be answered: How many different strains of poliovirus existed? How could a safe and steady supply of poliovirus be produced for use in a vaccine? And what, exactly, was the pathogenesis of the disease—how did it travel through the body and get to the central nervous system, where the damage occurred?

Question one would involve dozens of researchers and a vast pool of money. A successful vaccine had to protect against every strain of poliovirus, and nobody had a clue what that number might be. Finding out was tedious work, but absolutely essential. "I know of no problem in all the medical sciences that was more uninteresting to solve," Harry Weaver admitted. "[It meant] the monotonous repetition of exactly the same technical procedures on virus after virus, seven days a week, 52 weeks a year, for three solid years."

In all, 196 strains were tested, and all fit neatly into three distinct types. Unlike influenza (or HIV/AIDS to come) the poliovirus family proved remarkably stable and conveniently small. This was very good news, indeed.

Question two would be answered by John Enders, Fred Robbins, and Thomas Weller at the Children's Hospital of Boston. Their ability to culture poliovirus safely in animal tissue outside the body (in vitro) is considered one of the landmark achievements of modern virology. The three men, all March of Dimes grantees, would be awarded the Nobel Prize for Medicine—the only polio researchers to be so honored.

Question three would require the shared research of David Bodian at Johns Hopkins and Dorothy Horstmann at Yale, each with heavy March of Dimes funding. Previous theories had poliovirus entering through the nose and traveling directly to the brain and central nervous system without entering the bloodstream. What Bodian and Horstmann discovered is that poliovirus enters the mouth, passes through the digestive tract, and, in a small number of cases, does indeed move through the bloodstream into the nervous system. A vaccine producing antibodies in the blood could probably neutralize the virus.

Hundreds of scientists provided the building blocks for the life-saving vaccines developed by Jonas Salk and Albert Sabin. The beauty of the program overseen by O'Connor, Rivers, and Weaver was that it relied on the best available talent. At a time of intense anti-Semitism in the medical field, the two largest March of Dimes research grants went to Salk and Sabin—both Jews. At a time when prejudice against female scientists was commonplace, two of the most successful grants went to women—Isabel Morgan working on a killed-virus vaccine at Johns Hopkins, and Dorothy Horstmann at Yale. Pedigree meant little. Large sums went to researchers at Harvard, Hopkins, and Yale, but equally large amounts were given to researchers at less prestigious places. Quality alone determined these awards.

In 1954, the March of Dimes sponsored the largest public health experiment in American history—the field trial of Jonas Salk's killed-virus polio vaccine. More than a million school children were involved: one group receiving the real vaccine, a second group getting a look-alike placebo, and a third group acting as "observed controls." The experiment was double-blind, meaning that neither the child receiving the shot nor the person giving it knew what was in the needle. The vaccine, produced by Parke–Davis and Eli Lilly, had been triple tested for safety: first in Salk's laboratory, next by the manufacturers, and then by the National Institutes of Health—the only government involvement in the entire experiment.

Many in the scientific community were skeptical of Salk's vaccine. They viewed Salk as a relative novice, barely 40 and relentlessly ambitious, who had not paid his dues. They fretted that his killed-virus vaccine was not strong enough to produce lasting immunity or safe enough to be mass-tested on children. But O'Connor not only had faith in Salk, he also had the last word. Thousands were being struck down each year by polio, and Salk alone appeared to grasp the urgency of the moment. As O'Connor put it: "He sees beyond the microscope."

It would take a full year to tabulate the results. On April 12, 1955, as families huddled around radios, the verdict came down: Salk's vaccine was "safe, potent, and effective." At a White House ceremony honoring the achievement, President Dwight D. Eisenhower choked back tears as he told Salk and O'Connor: "I have no words to thank you. I am very, very happy."

Three years later, Albert Sabin successfully tested his live-virus oral polio vaccine in the Soviet Union and other Iron Curtain countries. By 1961, as Sabin's vaccine replaced Salk's, the number of polio cases in the United States dropped below 1,000 and would soon be in double digits, virtually wiping polio off the American map.

There was a problem, however. In an extremely small number of cases—about 1 in 750,000—the live virus in Sabin's attenuated vaccine reverted to virulence, causing polio in the child. As a result, polio could never be fully eradicated—a reality that everyone but Sabin himself sadly acknowledged. In the 1990s, the United States went back to a more effective version of the killed-virus Salk vaccine. Polio completely disappeared.

The enormous efforts of the March of Dimes in the domestic polio crusade have been taken up globally by groups like Rotary International, the Bill and Melinda Gates Foundation, the World Health Organization, and the Centers for Disease Control and Prevention (CDC). The aim is to make polio the second infectious disease in human history to be wiped off the face of the earth, smallpox being the first. And both vaccines—the killed-virus Salk and the live-virus Sabin—will be needed to finish the job. What greater tribute to the memory of Basil O'Connor than to see this noble goal fulfilled.

David Oshinsky, PhD
Director, Division of Medical Humanities
NYU School of Medicine
August 2015

ACKNOWLEDGMENTS

The philosopher and critic Walter Benjamin claimed that nothing that has taken place should be lost to history, but that only to redeemed humanity does the past belong in its entirety. Accordingly, these pages will show that Basil O'Connor has been pulled back into view from the prolonged neglect that precedes becoming lost, and that in his vision for the March of Dimes as a quest for "Freedom from Disease" he hungered to redeem humanity for future generations of children from the disasters of polio, birth defects, and premature birth. The endeavor of biographical and historical reconstruction is never an isolated one, and the following individuals and institutions contributed in various ways with vital assistance and spirited encouragement. It is a pleasure and an honor to acknowledge my debt of gratitude to all those who invigorated this project with their knowledge, expertise, and support.

The original research for this project was supported by a Lubin–Winant Research Fellowship from the Franklin and Eleanor Roosevelt Institute. On behalf of the March of Dimes Foundation, I express my heartfelt gratitude to the Institute for this generous gift of support. To the staff of the Franklin D. Roosevelt Presidential Library in Hyde Park, New York, I wish to express my enthusiastic thanks not only for the consummate professionalism and steadfast dedication to serving researchers from every walk of life but also to reflect their own surplus of Rooseveltian generosity back to them with the deepest appreciation. The FDR Library is an archival and historical treasure! In particular, Lynn Bassanese, Robert Clark, Clifford Laube, Franceska Macsali-Urbin, Sarah Malcolm, and Jeffrey Urbin have all provided invaluable assistance to the author, and the March of Dimes counts them as treasured friends and partners.

Angela N. H. Creager of Princeton University and Daniel Wilson of Muhlenberg College kindly reviewed a first draft of this writing, as did Charles Massey of the March of Dimes. Dr. Creager and Dr. Wilson provided helpful advice on many occasions as well as a sustained dialogue about the history of 20th-century science, medicine, and disease that proved invaluable. David Oshinsky of New York University and Peter Salk of the Jonas Salk Legacy Foundation did the same. Dr. Oshinsky's vivid presentations on the history of polio remain second to none; his assistance has been consistently generous, sparkling with insight; and we thank him for his beautifully written foreword to this volume and for his many years of support. Dr. Peter Salk has shared his personal knowledge of his father Jonas Salk and the history of the polio vaccine with exacting attention to the nuances of historical interpretation, providing many colorful details of his own experiences with Basil O'Connor. I am proud to say that Peter and I first met when we shared the podium at a commemoration of the 50th anniversary of the Salk polio vaccine announcement at NYU in 2005.

I extend my profound thanks to March of Dimes President Dr. Jennifer Howse and to Senior Vice President and Chief Medical Officer Dr. Edward McCabe for the opportunity to help preserve the legacy of Basil O'Connor and the March of Dimes in this monograph series. Dr. McCabe reviewed my writing with insightful criticism every step of the way, and I deeply appreciate his expert guidance and personal support. Many thanks to Senior Vice President of Strategic Marketing and Communications Doug Staples and the Media Relations Team of Michele Kling, Todd Dezen, and Elizabeth Lynch whose keen sense of history as a work in progress provides a constant learning experience. I am especially grateful to Michele for her fund of historical knowledge and insightful review of a first draft of my writing. I extend my sincere thanks also to Dr. Michael Katz and to Dr. Christopher Howson for their expert tutelage regarding the international dimensions of the Foundation and for sharing their wealth of knowledge of the history of science and medicine. Thanks to Cynthia Pellegrini of the Office of Advocacy and Government Affairs, and Beverly Robertson and Lilliam Acosta-Sanchez of the Pregnancy and Newborn Health Education Center, for helpful information on many occasions; to Motoko Oinuma for technical support; and to Peter Coletta whose knowledge of film history and technology is always most helpful. To my wife, Susan Rose, Director of Program Services, March of Dimes New York State Chapter, I offer my deep appreciation for her critical insights and supportive understanding throughout this project.

Many others have provided expert knowledge and helpful assistance on more occasions than I can tell: sincere thanks to Chuck Dittrich, Deborah Gardner, Ronald Green, Joan Headley, Caitlin Hawke, Julia Marino, Jan Nichols, Adam Pellegrini, Edith Powell, and Phil Schaap. Thanks to the members of the 21st century incarnation of the Cuff Links Club in their enthusiasm for FDR, in particular Dr. Steven Lomazow, Dr. Hal "Toby" Raper, Frank Costigliola, and Robert A. Friedman. Of the many scholars who have utilized the March of Dimes Archives, I am very grateful for a stimulating rapport with Richard Altenbaugh, Suzanne Bourgeois, William Cleveland, Charlotte Jacobs, Alex Kertzner, Stephen Mawdsley, Naomi Rogers, and Durahn Taylor.

A very special thanks to Cathy Hively, granddaughter of Basil O'Connor, for sharing her reminiscences about "Doc" and Hazel in lengthy interviews on several occasions, and for her warm friendship and generous support of the March of Dimes.

Thanks also to fellow archivists Michael Shadix, Roosevelt Warm Springs Institute for Rehabilitation; Peter Carini, Rauner Special Collections Library, Dartmouth College; Christine Beauregard, Manuscripts and Special Collections, NY State Library; and Susan Watson, Hazel Braugh Records Center, American Red Cross. My editors at Elsevier have been most helpful and efficient; they are Molly McLaughlin, Erin Hill-Parks, and Catherine A. van der Laan.

Finally, this project could not have been completed without the assistance of Saira Suri, an undergraduate attending Vanderbilt University and summer intern at the March of Dimes in 2013, 2014, and 2015. Saira volunteered her assistance with research, bibliography, footnotes, style, and organization. She also helped to organize two key collections of the March of Dimes Archives: the papers and records of Dr. Michael Katz and the Office of Government Affairs Records. Her help was timely and invaluable—thank you, Saira!

This volume is respectfully dedicated to the memory of Charles L. Massey, President of the March of Dimes from 1978 to 1989. Charlie loved the March of Dimes, and he further stimulated my exploration of its history. In 20 recorded interviews in 2004 and countless occasions afterward he related his personal experiences to me, an unforgettable journey which I will treasure always. As F. Scott Fitzgerald wrote in *The Great Gatsby*:

So we beat on, boats against the current, borne back ceaselessly into the past.

David W. Rose
Archivist, March of Dimes
September 2015

Peoples far beyond the frontiers of the United States rightly looked to him as the most genuine and unswerving spokesman of democracy of his time, the most contemporary, the most outward-looking, the boldest, most imaginative, most large-spirited, free from the obsessions of an inner life, with an unparalleled capacity for creating confidence in the power of his insight, his foresight, and his capacity genuinely to identify himself with the ideals of humble people.

Isaiah Berlin on Franklin Delano Roosevelt, Personal Impressions (1980)

He was different from anybody else in the crowd. He was very much in command of the situation. I don't know why—but the term comes to me because I've thought of him since then that way—but he was like the Medicis. He had it within his power to cause almost anything to happen. … He used his position and his power for a single-minded purpose, and he was sharp, intelligent: he was so scientific in his way of thinking that at times I thought that he put my colleagues to shame. I would say that the fact that the vaccine became available in 1955 was attributable to the existence of Basil O'Connor.

Jonas Salk on Basil O'Connor, September 8, 1984

CHAPTER 1

Introduction: The March of Dimes and Historiography

In his State of the Union address on January 20, 2015 President Barack Obama announced the launch of a new initiative that he called the "Precision Medicine Initiative" to refine and personalize medical treatment. The initiative is intended to develop a program to coordinate individual genetic, lifestyle, and symptomatological data to personalize diagnosis and treatment with the precision appropriate to specific characteristics of the human individual and not an "average" patient. President Obama stated, "I want the country that eliminated polio and mapped the human genome to lead a new era of medicine—one that delivers the right treatment at the right time."[1] His reference to these two accomplishments—the elimination of a devastating epidemic disease and one of the most prominent milestones in the science of genetics—was made without mention of the organization that played a significant role in the history of both. The March of Dimes, founded in 1938 as The National Foundation for Infantile Paralysis (NFIP), brought the awful tribulations of polio to a close in the United States just after the midcentury, only to renew its battle against disabling diseases by stimulating research into the causes of birth defects through genetics and genomics. The story of polio's demise is well known thanks to the dramatic aura that clings to the story of a robust politician stricken in the prime of his life and crippled by "infantile paralysis" who then went on to become president of the nation, and to the legacy of the organization that democratized philanthropy by involving ordinary people in the social construction of scientific accomplishment through a continuous multitude of small donations. Less appreciated, and frequently overlooked, in the history of the human genome project are the human gene mapping workshops that were sponsored by the March of Dimes that formed the incremental steps toward the ultimate goal of mapping the entire genome. In fact, Dr. Victor McKusick, fondly known as "the Father of Medical Genetics," put forward the very idea of the human genome project at a March of Dimes International Birth Defects Conference in 1969, a conceptualization that the March of Dimes

---

[1] *The Lancet.* "Planning for US Precision Medicine Initiative underway". 385:2448. June 20, 2015.

Friends and Partners
ISBN 978-0-12-803597-9
http://dx.doi.org/10.1016/B978-0-12-803597-9.00001-1

1

then translated into immediate action. President Obama's call for a new era of "precision medicine" invokes not only the history of the March of Dimes but also its capabilities concerning the efficacy in medicine as it pursues the ultimate goals of putting an end to birth defects and premature birth in our lifetime.

The place of the March of Dimes in American medical history tacitly invoked by President Obama holds a small irony in that accomplishments cited—polio eradication and the genome project—were spearheaded by a small, independent organization as opposed to a vast governmental bureaucracy of entangled departments and agencies. That the detailed history of the March of Dimes *as an organization* has not been a subject of focused historical research is the question set forth here. Founding by a major US President, eradicating polio, joining the aims of science with the lives of ordinary people, and reinventing itself as a leader in birth defects prevention—all of these are well known and individually treated in disparate media, in print, and online. The story of the eradication of polio as a threat to public health has been explored from several perspectives since it is a signal accomplishment that resonates powerfully in the 21st century with a world initiative on the verge of total global eradication, which involved many organizations and individuals in collaboration with the March of Dimes, without whom the Salk and Sabin polio vaccines may not have been developed so early (1955 and 1962, respectively). While incisive historical and biographical studies such as Daniel Wilson's *Living with Polio*, David Oshinsky's *Polio: An American Story*, and Charlotte Jacobs' *Jonas Salk: A Life* have contextualized the historical particulars of the polio epidemics and the vaccines as an *American* story, the organizational history of the March of Dimes has remained untouched and curiously out of focus, except insofar as acknowledged as the engine for success. And while the history of polio in America gained closure as an immediate problem, the reorganization of the March of Dimes in 1958 toward the complexities of its multidimensional attack on birth defects and premature birth are well known yet somehow appreciated by historians and film makers only piecemeal, still wanting proper historical synthesis.

This study does not correct this historical discrepancy in a comprehensive overhaul but hopefully provides a benchmark for future organizational history, one that is measured by its original source in historical biography. The friendship and partnership of Franklin D. Roosevelt (FDR) and Basil O'Connor remains central to the creation of the NFIP and its fund-raising campaign known as the "March of Dimes"; their personalities have inflected the history of polio prevention in ways that impart special character to their humanitarian collaboration.[2] Besides the enormous breadth of their individual accomplishments, each is a "colorful character" with an indelible impact on American culture, but this study is not a biography proper of either Roosevelt or O'Connor in the sense of a written portrait of a life as a conclusive summing up from birth to death. The field of Rooseveltian biography is ongoing and inexhaustible, to be referenced here primarily as it pertains to the history of his personal relationship with Basil O'Connor; while the portrait of Basil O'Connor here, though tending toward a more traditional biographical portraiture, is scrutinized largely in the circumstances of his association with FDR and the formation of the March of Dimes. Most essential are their interactions, each man's influence on the other, and the common ground of their careers. As Frank Costigliola has vividly demonstrated in his *Lost Alliances: How Personal Politics Helped Start the Cold War* (2011) interpersonal relationships and the fluctuating dynamics in the Roosevelt administration did indeed have an authentically world-changing impact in political history. Costigliola did not include Basil O'Connor in his examination of Roosevelt's ensemble of advisors in the monumental attempt to steer the Soviet dictator Josef Stalin toward mutuality, diplomacy, and tractability to American ends in the concluding year of World War II. However, FDR's lasting confidence in O'Connor made him privy to momentous secrets even in the arena of global geopolitics, far afield from the daily management of the NFIP. Long after FDR's death, O'Connor suggested that the President had revealed to him facts about the Yalta Conference the knowledge of which would not outlive either of them.[3]

The basis of the historiography of the March of Dimes actually begins with projects of FDR and Basil O'Connor themselves. FDR signed the legislation to create the National Archives and Records Administration (NARA)

---

[2] "The National Foundation for Infantile Paralysis," herein abbreviated "NFIP," was the legal, corporate name of the Foundation from 1938 to 1958, at which time the name was shortened to "The National Foundation" with the mission change to birth defects prevention. The "March of Dimes" was the name of the annual fund-raising campaign of the NFIP beginning in 1938, a designation popularly used to refer to the Foundation itself through its history. The Foundation changed its corporate name to the hyphenated form "National Foundation-March of Dimes" in 1967, to "The March of Dimes Birth Defects Foundation" in 1979, and finally to "March of Dimes Foundation" in 2006. In general, "March of Dimes" may be loosely (and correctly) used to refer to the Foundation through its entire history; however, "The National Foundation for Infantile Paralysis" or "NFIP" will more accurately refer to the Foundation in its first 20 years. This usage is preferred here.

[3] Costigliola, Frank. Roosevelt's Lost Alliances: How Personal Politics Helped Start the Cold War. Princeton: Princeton University Press, 2012.

and created the first presidential library, a model followed by all subsequent American presidents. In the course of launching his library, FDR brought Basil O'Connor directly into the endeavor of preserving his presidential and personal papers. O'Connor thus became involved in the establishment of the FDR Library, and later the Harry S. Truman Library as well. These formative and highly visible roles strengthened O'Connor's already acute historical consciousness, which he brought into play during the years of the Salk polio vaccine field trial in the mid-1950s to preserve the history of the NFIP, formally and professionally. Perhaps amplified by a vain sense of his own historical significance, O'Connor retained the management consulting firm Hackemann & Associates of Madison, Wisconsin in 1953 to research and write a formal history. For four years, Louis Hackemann and his associate Ruth Walrad, authors of a history of the American Red Cross whom O'Connor had met during his stint as president of that organization, orchestrated the project through a newly created Historical Division of the NFIP to chronicle its achievements in its minutest details. The result was a multivolume, 3,000-page account of the NFIP in its several fields of significant endeavor: national administration and policy, fund-raising, medical research, medical services, professional education, and public education. The study encompasses the activities of the Georgia Warm Springs Foundation from the mid-1920s, the emergence of the NFIP, and the first efforts toward a comprehensive program to stop the polio epidemics in the United States. A separate volume on the Salk polio vaccine field trial was planned but never completed. O'Connor and the NFIP sought a publisher at the completion of the project in 1957 but with no success. The unpublished study remains a cornerstone of the collections of the March of Dimes Archives, providing documentation of the earliest years of the foundation and relying entirely on original materials available at the time of writing, some of which are regrettably no longer extant. The *History of the National Foundation for Infantile Paralysis* remains a singular but ponderous account of a noble experiment in disease prevention that resonated from the presidency to the science establishment through Hollywood and the mass media and that colored American culture and society in ways that were unique in the American experience.[4]

---

[4] March of Dimes Archives (hereafter abbreviated "MDA"), The History of the National Foundation for Infantile Paralysis Records, 1936–1993. This collection consists of the final draft typescript titled *History of the National Foundation for Infantile Paralysis*, preliminary drafts of each chapter, and related supplementary materials.

Angela Creager of Princeton University was the first contemporary historian to utilize this massive historical study in her research, an examination of the work of Wendell Stanley (a grantee of the NFIP) and his path-breaking work on tobacco mosaic virus in *The Life of a Virus: Tobacco Mosaic Virus as an Experimental Model, 1930–1965* (2002). Several others have followed her lead. The future historian considering an organizational or narrative history of the March of Dimes will have additional resources: not only the plentitude of original source materials of the foundation's Archives, but other important publications as well, including the published proceedings of the International Poliomyelitis Conferences (1948–60) followed by the International Birth Defects Conferences (1960–77); the *Birth Defects: Original Article Series* (1965–95) described below; and the public opinion surveys conducted in the mid-1950s that led to the publication of *The Volunteers: Means and Ends in a National Organization* (1957), by David Sills, which was the first sociological assessment of the NFIP as a voluntary organization. Once again, it was Basil O'Connor who set all of these projects in motion. The breadth of O'Connor's intellectual compass as a visionary leader soars far beyond the reductive simplicity of the common stereotype of the man as the head of the NFIP.

Distinctly aware of his own role in the very thick of "history-in-the-making" during the planning of the field trial of polio vaccine developed by Dr. Jonas Salk in 1954, a clinical trial touted by the March of Dimes as the "largest peacetime mobilization of volunteers" in American history, O'Connor also mandated the requirements for change. By 1953 he completely trusted Salk's vaccine research as the solution to the problem of polio and recruited a former Red Cross associate, Melvin Glasser, to investigate the best opportunities for the future direction of the organization. Thus, just as the Salk vaccine was being developed and tested, the NFIP began a comprehensive analysis of its impact in the field of public health through formal surveys conducted by the Bureau of Applied Social Research and by the American Institute of Public Opinion. These studies solidly confirmed the organizational strength and broad base of public support for the medical program of the NFIP that helped to shape the foundation's new mission against birth defects in 1958 with the launch of its Expanded Program. According to a Board of Trustees memo, the Gallup and Columbia University polls when completed "...were designed to determine what endeavor, if any, the National

Foundation should follow when the problem of poliomyelitis had been completely solved."[5] The privately commissioned Columbia University study, *The National Foundation: Its Volunteers and Public Support* (1954), which was the first survey of volunteers and supporters of the NFIP, was recapitulated in David Sills' *The Volunteers*, an examination of the March of Dimes as a social movement, its organizational structure, the social origins of local chapters, and the nature of voluntarism as a social experiment. The contemporaneous study of the NFIP by George Gallup's American Institute of Public Opinion also facilitated decision making about the mission change to birth defects. Gallup himself became a member of the NFIP board of trustees in the 1950s, undoubtedly favored by Basil O'Connor's appreciation of Gallup's famous success in predicting FDR's reelection in the 1936 presidential campaign. The March of Dimes has returned to the Gallup poll on many occasions to survey public opinion and attitudes about its programs and initiatives, for example, the National Folic Acid Campaign of 1998 to 2002. In sum, the two surveys, along with Melvin Glasser's transformational analysis of organization change and Dr. Josef Warkany's prodding O'Connor with great urgency to choose birth defects prevention as a new mission, all led to the announcement of the so-called Expanded Program. O'Connor, Glasser, and their advisors reached the conclusion that prevention of birth defects and arthritis (which was discontinued in 1963) were achievable mission objectives not because they were "unmet medical needs" but that the NFIP was equipped to meet these needs, as stated explicitly in the Columbia University survey.[6] At the same time, the Foundation simplified its name from the NFIP to "the National Foundation." Basil O'Connor's continuing confrontation with a widening spectrum of disease became apparent in a new direction: he idealized the foundation as a "flexible force" in the field of public health, an innovative twist in the Rooseveltian notion of "Freedom from Disease."

The organization of the NFIP as a national civic movement based on a confederated network of local cadres of volunteers gave way to an even more structured hierarchy after the polio years, just as Josef Warkany's teratological model of birth defects gave way to an understanding of "congenital malformations" based on genetics. The 1960s were not only years of change and reorganization, but also of summing up, with polio a

[5] MDA, Board of Trustees Records. No. 6, June 18, 1954.
[6] MDA, Medical Program Records. Series 6: Expanded Program, "Development and Planning, 1952–1959." October 8, 1959.

lesser threat than it had been at the founding of the NFIP. Saul Benison's oral history of virologist Thomas Rivers is the preeminent study in a second wave of historical reconnaissance of how polio came to be defeated. *Tom Rivers: Reflections on a Life in Medicine and Science* (1967) is an "oral history memoir" and as such is a classic model of its kind. In this memoir, Benison achieves an unparalleled, wide-ranging history of mid-20th-century virology and medicine expressed in the personal reflections of Dr. Tom Rivers, a brilliant but truly cantankerous and feisty decision maker in the momentous events leading up to the Salk vaccine field trial and its aftermath (not to mention NFIP grant making). Thomas Milton Rivers, MD (1887–1963) was an American origin, an intellectual powerhouse of the kind that Basil O'Connor needed most to fulfill his vision of the "precision medicine" of his time. Rivers belonged to an age of medicine that knew viruses as "filterable viruses" (because they traveled freely through porcelain filters that otherwise easily trapped bacteria), and he led the Rockefeller Institute as Director when he joined the NFIP. His important role as a leader in NFIP grants administration and key player in planning the 1954 field trial are revealed in painstaking detail in Benison's oral history.

Saul Benison was one of the original members of the Oral History Research Office of Columbia University with historian Allan Nevins, who founded the office in 1948. Benison was as close as anyone to the professional intellectual source of oral history in the United States, later publishing "Reflections on Oral History" in the *American Archivist* and acknowledging Nevin's role in the field with a dedication in his Tom Rivers memoir: "For Professor Allan Nevins/Known to my generation at Columbia as the Little Shepherd/From one of the sheep."[7] Benison became Professor of History at Brandeis University until his retirement in 1990, and he benefitted from a warm relationship with Basil O'Connor, becoming more or less official historian of the National Foundation in the 1960s. Before any other scholar, Benison recognized the world historical significance of the NFIP in the milieu of global war when FDR was yet alive, expressed in a stunning observation that has continuing relevance for the March of Dimes mission in both national and international contexts today. In an unpublished essay, "The National

---

[7] Benison, Saul. "Reflections on Oral History," *The American Archivist*, 28(1):71–77. January 1965. See also *Tom Rivers: Reflections on a Life in Medicine and Science*. Cambridge (MA), Massachusetts Institute of Technology University Press, 1967.

Foundation: 1938–68, A Retrospective View," later reprised in an article on the history of polio research in the United States, Benison stated:

> By 1936 President Roosevelt's conviction grew that polio could only be conquered through a broad and sustained program of scientific education and research. The organization of The National Foundation for Infantile Paralysis was in essence the first step toward the realization of that goal. The new voluntary health organization was also something more. At a time when deadly assaults had already been launched against the human spirit and life itself in Europe, the new Foundation stood as an affirmation of the value of conserving human life and dignity. Ordinary people everywhere recognized this quality and quietly and emphatically made the cause of the new Foundation their own. That support never faltered.[8]

Benison also authored several articles on the history of polio in addition to the Tom Rivers memoir and an unpublished "retrospective view" of the National Foundation's first 30 years; and he collected and indexed Basil O'Connor's speeches, having unique access to an earlier incarnation of the March of Dimes Archives than the existing archival collections of the March of Dimes today. Benison's meticulous notes on manuscript and bibliographic sources for the Rivers memoir reveal an extensive archival collection of documents, indisputably unique for a private nonprofit foundation of the 1960s, which he describes in part as follows:

> The Archives, which are kept at the National Foundation's headquarters in New York, contain approximately 3,000 linear feet of records, organized into seven major record groups, including legal records, accounting and financial records, medical service records of patients supported by the Foundation, chapter administrative records, administrative records of the national office of the Foundation, public relations records, and grant file records.[9]

Many of these core collections are still preserved and accessible today in the contemporary March of Dimes Archives, but the discrepancies between the current collection and Benison's bibliographic snapshot of 1967 leads to the disappointing conclusion that much has been lost over the years. While Benison's work remains fundamentally important, two other books of that era deserve acknowledgment as participating in the second wave of historical retrospectives on polio and the March of Dimes. John Paul's *History of*

---

[8] Benison, Saul. "The National Foundation: 1938–1968, A Retrospective View," unpublished typescript in MDA, Medical Program Records. Series 14: Poliomyelitis. Benison, Saul "Typescripts." See also Saul Benison, "The History of Polio Research in the United States: Appraisal and Lessons" in Holton, Gerald, editor. *The Twentieth Century Sciences: Studies in the Biography of Ideas.* New York: Norton, 1972.

[9] Benison, Saul. *Tom Rivers: A Life in Medicine and Science.* Cambridge (MA): Massachusetts Institute of Technology University Press, 1967, p. 623.

Poliomyelitis (1971) is an authoritative account of "the natural history" of this disease, based on Dr. Paul's long experience as a founding investigator, with James Trask, of the Yale Poliomyelitis Study Unit and a research grantee (the very first) of the NFIP. Compared with Benison's meticulous citation of sources, Paul's history lacks such scholarly apparatus; instead, he relied on Benison himself for assistance with areas of history beyond his own direct experience as a polio epidemiologist beginning in the 1930s. Dr. Paul likened the appearance of the NFIP to "the sudden appearance of a fairy godmother of quite mammoth proportions who thrived on publicity," lending to the polio literature a notorious characterization that has been recycled ad nauseam.[10] Finally, Richard Carter's *Breakthrough: The Saga of Jonas Salk* (1966) is a journalistic attempt to portray the public drama of the Salk vaccine, and though he benefitted from interviews with Basil O'Connor as well as the assistance of Saul Benison and others close to the story of the vaccine, it is undistinguished by the meticulous objectivity in Benison's own approach to the same events in his life and of Tom Rivers. Stanley Plotkin, MD, then of the Wistar Institute, who developed the RA27/3 rubella vaccine, pinpointed the rudiments of a simplistic conspiracy theory in *Breakthrough*, finding a drastic imbalance of perspective in Carter's favoritism of Jonas Salk as the only polio researcher among his peers who was objective and altruistic. In a book review in *Clinical Pediatrics*, Dr. Plotkin wasted no words in stating right off: "This is essentially a paranoid book, but like many paranoid productions not without merit or intrinsic truth."[11] Richard Carter's *Breakthrough*, written much closer in time to the events it portrays, has been superseded by other fine histories that utilize original source materials rather than an obsolete old boy network of cronies and newsmen.

In 1965, under the editorial leadership of Daniel Bergsma, MD, the March of Dimes launched the *Birth Defects: Original Article Series* (BDOAS), a periodical publication that became a key component of the foundation's growing professional education program. Dr. Bergsma was Director of Professional Education from 1959 to 1977 at a time when the foundation's corporate name was hyphenated as The National Foundation-March of Dimes. He also authored the *Birth Defects Atlas and Compendium* (1973), a massive encyclopedia of malformation syndromes and genetic

---

[10] Paul, John R. *A History of Poliomyelitis*. New Haven (CT): Yale University Press, 1971, p. 311.

[11] Plotkin, Stanley A. Review of *Breakthrough: The Saga of Jonas Salk*. *Clinical Pediatrics*. October 1966, p. 581.

diseases. The BDOAS was created "to enhance medical communication in the birth defects field," and it published original papers on inherited disorders, genetic counseling, immunology, embryonic development, and bioethics by medical specialists in genetics and birth defects research. Also featured were the papers and proceedings of medical symposia sponsored or cosponsored by the March of Dimes beginning with the Symposium on the Placenta in 1965, edited by Dr. Bergsma himself. Robert Good, MD coedited volumes on immunodeficiency diseases as did Victor McKusick, MD with several volumes on the clinical delineation of birth defects. Frank H. Ruddle, MD was involved with Drs. McKusick, Bergsma, and others in the publication of nine international human gene mapping workshops sponsored by the March of Dimes. The BDOAS, along with the proceedings of the International Birth Defects Conferences and the annual March of Dimes Conference on the Clinical Delineation of Birth Defects established in 1968 in association with the Johns Hopkins Medical Institutions, were cornerstones documenting an enormous transformation: the systematic and holistic turn to birth defects prevention. Victor McKusick, chair of the first five of the clinical delineation conferences, equated the study of birth defects with medical genetics, believing that the field was "an effective focus for the March of Dimes, clarifying its mission as a way to unravel the basic problems [of genetics]."[12] Dr. McKusick was instrumental in developing the annual Short Course on Medical and Experimental Mammalian Genetics with March of Dimes support at the Jackson Laboratory in Bar Harbor, Maine, claiming of the March of Dimes role in medical genetics that "*massive* might not be an excessive evaluation."[13] In sum, all of these publications and conference proceedings originating in the 1960s and continuing to the turn of the millennium are published resources to be utilized in conjunction with original source materials of the March of Dimes Archives in future historiography of the foundation.

A resurgence of interest in oral history in the 1980s anticipated the 50th anniversary of the March of Dimes in 1988 with a project "to create an anecdotal account of the March of Dimes based on the personal remembrances of people who had shaped its history."[14] The project was carried out by Albert Rosenfeld, Consultant on Future Programs, and Gabriel Stickle, who served as Statistician for the Salk polio vaccine field

---

[12] McKusick, Victor. Personal interview. October 22, 2007.
[13] *Ibid.*
[14] MDA, Oral History Records, 1980–1992.

trial and later became Executive Assistant to the President, Basil O'Connor, in 1966. The transcribed tape recordings of 57 interviews conducted from 1983 to 1988 were never published, though the reflections of many important figures in March of Dimes history, namely Jonas Salk, Victor McKusick, Elaine Whitelaw, Dorothy Ducas, and other volunteers and staff, were documented. In the meantime, renewed interest in the earlier polio mission of the NFIP was brewing in the 1990s long after the foundation had broadened its birth defects mission to encompass the field of perinatology and the problem of premature birth. Filmmaker Nina Gilden Seavey utilized the March of Dimes audiovisual collection for research on the history of polio; her award-winning documentary, *A Paralyzing Fear: The Story of Polio in America*, was released in 1998, a stimulus to the founding of the March of Dimes Archives as an institutional repository for the original records of the foundation. Since 1998, a spate of original studies—a third wave of historical reconnaissance—have utilized this collection: *The Life of a Virus: Tobacco Mosaic Virus as an Experimental Model, 1930–1965* (2002) by Angela N. H. Creager; *Living with Polio: The Epidemic and Its Survivors* (2005) by Daniel J. Wilson; *Polio: An American Story* (2005) by David Oshinsky; *The Cutter Incident* (2005) by Paul A. Offit, MD; *Bracing Accounts: The Literature and Culture of Polio in Postwar America* (2008) by Jacqueline Foertsch; *The Polio Years in Texas: Battling a Terrifying Unknown* (2009) by Heather Green Wooten; *The Immortal Life of Henrietta Lacks* (2010) by Rebecca Skloot; *Genesis of the Salk Institute: The Epic of Its Founders* (2013) by Suzanne Bourgeois; *Polio Wars: Sister Kenny and the Golden Age of American Medicine* (2014) by Naomi Rogers; *Jonas Salk: A Life* (2015) by Charlotte Jacobs; and *Selling Science: Polio and the Promise of Gamma Globulin* (2016) by Stephen Mawdsley.

Any of these studies may be destined to become classics in the medical humanities. One of them, *Polio: An American Story*, has achieved acclaim with a Pulitzer Prize in History in 2006. These and other published articles, dissertations, films, and exhibits amount to a minirenaissance in the discovery that the March of Dimes exists in the firmament of Americana as well as in traditional historiography of science and medicine. And yet there is no institutional history of the March of Dimes as an organization, nor do any of these studies examine the history of the foundation (or even the polio years) from an organizational perspective in a sustained fashion. The reasons for this seeming neglect are paradoxically obscured by the very fact that so many have taken stock of the foundation's singular achievements in so many areas of endeavor. The complexity and scope of this history is daunting, and the Foundation naturally moves ahead without looking back nostalgically even as it deploys its history of achievement strategically to impel its current mission. The mythologizing of the founder FDR is another factor, but historians are normally adept at decoding any institutional "myth of origins" that smothers the clarity of fact in the fogbank of legend. FDR's partner Basil O'Connor may be partly responsible as well, which is the crowning irony. O'Connor's heightened sense of himself as a historical actor on the global stage contradicts his failure to publish *The History of the National Foundation for Infantile Paralysis* in 1957 just as his role in the preservation of presidential papers contradicts the sad lapse in preserving his own. His public persona belies his reclusive method. And though public controversy attracts attention and makes for good theater, O'Connor's combativeness in the daily newspapers (eg, with Albert Sabin) tinges the reputation of his final years with an element of unbending stodginess that blends perfectly with the visible trappings of privilege and high style that he flaunted. His personal success was envied, or not understood in relation to his humanitarianism and liberal politics (yet another contradiction), and distortions of interpretation occurred much later in some attempts to portray O'Connor and the NFIP with any semblance of accuracy. One of these relates to the social construction of fear.

The idea that March of Dimes publicity was grounded in fear-mongering, that it preyed on the fears of a vulnerable public, that its success was predicated on fear-based marketing is an unfortunate misconception that demands correction. David Oshinsky alludes to this maladroit gloss on the success of the NFIP in *Polio: An American Story*, but it crops up time and again with virtually no analysis or substantiation. *Paralyzed with Fear: The Story of Polio* (2013) by Gareth Williams exemplifies an extreme version of this ill-founded opinion. A gossipy medical history of polio centering on Great Britain (Williams is Professor of Medicine at the University of Bristol) *Paralyzed with Fear* revels in breezy language that throws any pretention to care about scholarship completely out of kilter: the NFIP is dismissed as a tentacular monopoly fanning the flames of polio hysteria with no standard of accountability; Jonas Salk is accused of pathological "tunnel vision," and Basil O'Connor is branded as a manipulative "tough-talking, short-fused bully." Citing *Patenting the Sun: Polio and the Salk Vaccine*, an earlier telling of the Salk story by Jane Smith,

Williams misrepresents the NFIP as "slick, arrogant fear-mongers, raising money through a campaign of terror." The NFIP "campaign of terror," he reiterates, "was a classic stick-and-carrot strategy, but with both elements carved from fear" and led by a "thick-skinned, abrasive pirate," a "pushy businessman." Williams even questions FDR's principles: "It is ironic that Roosevelt himself was instrumental in the campaign of terror which deliberately denied Americans [Freedom from Fear], and that he did not live to reassure himself that the end justified the means." Stereotypical characterizations such as these are inexcusable substitutes for careful research and objective presentation; these are untenable, unwarranted, and patently unfair.[15]

Then why does this misperception of the NFIP as "fear-monger" persist? The answer lies in the distortions that arose from its own success with polio, its transformation and rebranding in 1958, and its ability to evolve with the evolution of the news media. Indeed, a history of the March of Dimes could be written from the perspective of its creative engagement with communications media, harkening back to FDR's "fireside chats" on radio (a brilliant exercise in *allaying* public fears). As Public Relations Director, Dorothy Ducas pointed out that the NFIP had utilized the news media unlike any other public health organization before it, and she, rather than Basil O'Connor or anyone else, was responsible for the engagement of its PR Department with the mass media in the 1940s and 1950s. She stated bluntly: "I ran it like a newspaper or a news service and nobody else had done that."[16] Moreover, Ducas insisted on disseminating *dependable* information about polio through all available news outlets throughout the year, not just during March of Dimes campaign drives, to keep advances over polio in the public eye and to involve volunteers effectively in solutions to a common, horrendous, national problem. Gareth Williams' overstatement that there was "no escape from March of Dimes propaganda, as the NFIP's well-funded and well-connected press office exploited all the media to hammer home the horrors of polio," entirely disregards the abundant evidence of the prior public alarm over polio-paralyzed children during the first decade or more of the existence of the NFIP.[17]

[15] Williams, Gareth. *Paralyzed by Fear: The Story of Polio*. London: Palgrave Macmillan, 2013. pp. 122–24; 132–37, 192.

[16] MDA, Oral History Records. Dorothy Ducas interview, March 26, 1984.

[17] Williams, Gareth. *Ibid*, p. 133.

One of Them Had Polio, 1949. The NFIP and the Museum of Modern Art in New York sponsored a polio poster contest in 1949, a challenge to artists to represent visually the fight against polio. Herbert Matter, an advertising designer and photographer known for his pioneering use of photomontage in commercial art, won First Prize for his poster "One of Them Had Polio." Matter's son, whose image appears on the right, was a polio survivor. The message "skilled teamwork brought recovery" characterizes the polio prevention program of the NFIP. The poster was reproduced on its annual report of 1949, a year of serious polio epidemics throughout the nation.

Another element in this mischaracterization adds a touch of the bizarre to the reputation of NFIP productions: the film *In Daily Battle* (1947), produced by RKO Pathé, Inc. for the NFIP as an educational film about services available to communities in the grip of polio epidemics, *In Daily Battle* is better known as "The Crippler," a name that was a common trope for polio itself. In the film, a short introduction features US Surgeon General Thomas Parran championing the NFIP as a "volunteer army" promoting "the Fifth Freedom— freedom from disease." The next brief snippet is notoriously famous: a specter of polio emerges in the form of a child's shadow with a crutch, chuckling with menace *I am Virus Poliomyelitis* as it roams city and countryside bringing paralysis to all it touches. The remainder of the film portrays the efficient charity of a March of Dimes chapter providing financial aid to families touched by polio, a straightforward scenario with actress Nancy Davis (later First Lady Nancy Reagan) in the role of a prim March of Dimes chapter assistant: "Miss Manning—she represents *you!*" The message was explicitly clear: "to replace fear with preparedness."[18] NFIP interventions in the epidemics in Hickory, North Carolina (1944) and Jackson, Mississippi (1946) are also depicted, but the emphatic lesson was one of reassurance and dependable care. Even though *In Daily Battle* falls in line with many educational films that boosted the good work of chapter volunteers and progress over polio, the "crippler" episode has been repeatedly taken out of context as a "scare tactic." NFIP staff recognized this tinge of grotesquerie and cautioned that the film might not be appropriate for schoolchildren though the film ultimately reached audiences in both schools and movie theaters. One scholar has recently examined this film in the context of the Frankenstein mythos.[19] But just as the contemporary March of Dimes approaches the problem of prematurity from every necessary medical, social, and epidemiological perspective, so too did the NFIP treat the problem of polio holistically. One constant consideration of local chapters operating in the public interface to address polio spreading out of control was precisely the reactions of panic and fear.

From the great epidemic of 1916 and the days after FDR was stricken, fear of polio as a parental and public concern was immediate and real, but there were other fears as ubiquitous. Ira Katznelson, in *Fear Itself: The New Deal and the Origins of Our Time* (2013) has mused that a generalized fear of the future has never been laid to rest since the upheavals of the Great Depression and World War II. Katznelson identifies the roots of three most

---

[18] MDA, Film and Videotape Collection, *In Daily Battle*. 1947.

[19] Codr, Dwight. "Arresting Monstrosity: Polio, Frankenstein, and the Horror Film." *Publications of the Modern Language Association*, 129.2: 171–87.

prominent fears in the era of the New Deal (which he sees running from Roosevelt's inauguration in 1933 until the end of Harry Truman's term in office): the spread of totalitarian dictatorships, warfare and global violence as a permanent condition, and the racist structure of American society.[20] Add to those the specific dread of nuclear warfare since Hiroshima and the political paranoia of the Joseph McCarthy debacle, and one can begin to contextualize the fear of polio properly in the complexities of the age. While the NFIP was founded not to address these overarching geopolitical and structural problems, it has produced a general lesson that has relevance in every field of human endeavor: to replace fear with action. This was, par excellence, the lesson of FDR's first inaugural address and Basil O'Connor's subsequent modus operandi for the NFIP. It was the principle from which O'Connor derived an important corollary to battle polio at the community level: "what have you got and what do you need?" That is to say, what medical resources does the community already have in place, and what does it need from the NFIP to quell an epidemic. This instant form of needs assessment was buttressed by the mass distribution of public health literature designed to explain polio to allay fear, the training of teams of Polio Emergency Volunteers (PEVs) to aid communities in emergencies, and the creation of a nationwide structure of volunteers who organized polio relief and fund-raising in a dependable way year after year. In this way, the NFIP acted as an agent of social change as it formed a moral identity for the organization that in truth expressed the collective identities of hundreds of thousands of volunteers.

In his *History of Poliomyelitis*, John Paul saw Basil O'Connor as a leader not only of good causes, but of great ones. Paul affirmed that O'Connor "had that loyalty for the good cause that goes with militant magnanimity."[21] The militancy of his magnanimity might be expressed with peremptory impatience with his staff or with the soothing tones of an orator to win over a public audience. It was this Janus-faced persona that has provoked mis-judgments that he was little more than a "short-fused bully." Even though Charles Massey, third President of the March of Dimes, admitted that O'Connor sometimes betrayed an "explosive temper," this was but the irascible impatience by which he confirmed his "militant magnanimity." A leader need not always project the soft-spoken humility of a Gandhi to be genuinely humanitarian, but the effectiveness of his administration will always be judged by its resulting achievements. As Massey explained:

[20] Katznelson, Ira. *Fear Itself: The New Deal and the Origins of Our Time.* New York: Liveright, 2014. pp. 12–14.
[21] Paul, John. *Ibid.*, pp. 308–309.

*A guiding principle of the foundation in all of its activities was **the pursuit of excellence**, regardless of politics or parochial interests. That's why O'Connor relied on the unbiased recommendations of prestigious medical and research advisory committees. No matter what he thought of Sister Kenny, when his advisors recommended a grant to her organization, he did not stand in the way. It was the same later on, during the mad race to develop a polio vaccine. Although Dr. Salk at the University of Pittsburgh was well ahead, the foundation gave his rival, Dr. Sabin, full support of his laboratory at the University of Cincinnati. Even after Sabin complained that Salk had received favored treatment the foundation **continued** to give Sabin all he asked for.*[22]

Misrepresentations of character aside, there is also the resounding silence of utter neglect: Basil O'Connor has become the "forgotten man" (a phrase that FDR had used in a fireside chat). This is conspicuously evident in recent cinematic treatments of FDR's struggle with polio, where historical completism is forfeited in preference of dramatic effect. In the HBO television production *Warm Springs* (2005), FDR's associates Louis McHenry Howe and Missy LeHand appear on the sidelines, but O'Connor is absent entirely. He does not appear in the frivolous comedy-drama *Hyde Park on Hudson* (2012) starring Bill Murray as FDR, and in the epic Ken Burns film documentary *The Roosevelts—An Intimate History* (2014), neither O'Connor nor the NFIP is even mentioned in an otherwise excellent portrayal of Roosevelt's struggle with polio by biographer Geoffrey Ward. In *Polio and Its Aftermath* (2005), Harvard literary critic and polio survivor Marc Shell has asked, "What happened to the NFIP itself? Beginning in the 1950s, the NFIP as Roosevelt conceived it—as a real instrument of national health policy and practice—simply disappeared." Professor Shell concluded, "The bold experiment in medicine and public health pioneered by Roosevelt … was all but forgotten."[23] Professor Shell did not dig deeply enough. What actually happened was that the NFIP *as FDR and Basil O'Connor first conceived it* reinvented itself and continued with remarkable progress as a leader against birth defects in the 1960s toward its campaign to prevent premature birth in the 21st century. While the pages that follow are not intended to explore the full scope of the history of the NFIP in its finest detail, this monograph will hopefully bring a corrective understanding to lapses and misconceptions described here by examining the life of Basil O'Connor and his connections to FDR with a cold but sympathetic eye in order to situate him in his rightful place for more encompassing projects of historiography to come.

---

[22] Charles Massey, personal interview. July 7, 2004.

[23] Shell, Marc. *Polio and Its Aftermath: The Paralysis of Culture.* Harvard University Press, 2005. pp. 184–85.

CHAPTER 2

Behind the White Carnation: The Leadership of Basil O'Connor

Among the multitude of Franklin D. Roosevelt's (FDR) enduring accomplishments was the creation of the March of Dimes, founded as The National Foundation for Infantile Paralysis (NFIP) in 1938. FDR first issued a presidential proclamation for the creation of the NFIP on September 23, 1937 to "lead, direct, and unify" the fight against polio. The new foundation quickly became a beloved American institution through its popular March of Dimes fund-raising campaigns, and through which it claimed the conquest of polio with the Salk vaccine licensed on the 10th anniversary of FDR's death, April 12, 1955. In 1958, with polio on the wane, the foundation reinvented itself by launching an adventurous program directed at birth defects. As the birth defects program developed through the 1960s, the foundation's mission incorporated by necessity the burgeoning fields of genetics and perinatology, evolving into its present mission, strongly characterized by a campaign against premature birth launched in 2003, and still including the prevention of birth defects and infant mortality. These later developments, which FDR could not have foreseen, continue and thrive as part of his permanent legacy.

While the story of the first March of Dimes radio campaign in 1938 that deluged the White House in dime donations has often been told, the person who orchestrated these campaigns that revolutionized volunteerism and fund-raising in the United States has been nearly lost to history. Basil O'Connor (1892–1972), FDR's law partner of the 1920s, who played key roles in the Georgia Warm Springs Foundation (GWSF) and the President's "Brain Trust" in the 1930s, led the NFIP as president for 34 years, from 1938 to 1972. During that period he served, simultaneously, as head of the American National Red Cross (1944–49) and was instrumental in creating the Salk Institute for Biological Studies in 1960. Roosevelt and O'Connor are the only two nonmedical honorees in the Polio Hall of Fame at Warm Springs, Georgia; yet there is no complete biographical treatment of O'Connor's life that can validate his historical significance as "the architect in the fight against polio."

Friends and Partners
ISBN 978-0-12-803597-9
http://dx.doi.org/10.1016/B978-0-12-803597-9.00002-3

In actuality, Roosevelt and O'Connor together set the groundwork for the massive campaign to end the polio epidemics in the United States through the GWSF and NFIP. O'Connor's initial reluctance to create a polio center at Warm Springs has been noted, but his loyalty to FDR superseded his initial hesitation. Eventually, O'Connor wholeheartedly supported FDR's interest in the venture, becoming a passionate spokesman for the polio cause. Their relationship matured not only through the mutual interest in Warm Springs, but also through the legal, personal, and political advice that O'Connor provided to FDR as Governor of New York (1928–32) and President of the United States (1933–45). Although FDR served as titular head of GWSF until his death in 1945, often spending time at his home there known as the "Little White House," it was O'Connor who directed the foundation during the Roosevelt presidency, serving as treasurer and chairman of its Executive Committee, ultimately succeeding FDR as president. O'Connor's qualifications as the executive head of both the GWSF and the NFIP were impeccable, though he disingenuously claimed he was just a "dumb lawyer" and not a scientist. This self-deprecation scarcely concealed his strategy of surrounding himself with the finest medical, scientific, and management advisors of the time. His insistence on careful deliberation of "the facts" in any decision, even the most momentous, left no doubt among his peers of his keen abilities as decision maker when discussion was ended and all facts and advice were on the table.

The NFIP was officially organized on January 3, 1938 in part to establish a nonpartisan basis for polio fund-raising and research beyond the confines of Warm Springs. Realizing that the problem of polio could not be effectively controlled by a local institution, FDR's proclamation of September 23, 1937 focused on the need to create a national organization to address the problem of polio in a holistic fashion. He stated that the GWSF had devoted its efforts "almost entirely to the study of improved treatment of the aftereffects of the illness" whereas the new foundation would attack "every phase of this sickness."[1] O'Connor assumed responsibility for leading the NFIP as he continued to manage the program at Warm Springs.

[1] Franklin D. Roosevelt. Statement on the New National Foundation for Infantile Paralysis. September 23, 1937. The American Presidency Project. http://www.presidency.ucsb.edu/ws/?pid=15463. See also The National Foundation for Infantile Paralysis. Annual Report, 1939.

After Roosevelt's death on April 12, 1945, O'Connor succeeded him as president of the GWSF, headed the Roosevelt Memorial Commission, and continued to lead both the Red Cross (until 1949) and the NFIP (until his death in 1972). O'Connor led the Roosevelt Memorial Postage Stamp ceremony at Warm Springs in August 1945 with a speech, "Nothing Could Conquer Him," a title which summarized his undying admiration of FDR as statesman and polio fighter. Roosevelt's influence on the history of disability, polio, and the independent living movement has been repeatedly demonstrated. Through O'Connor, the March of Dimes embraced FDR's legacy and actively promoted its mission in the aura of FDR's colossal reputation as a president whose experience with polio led to the eradication of the disease entirely from the United States. O'Connor's loyalty to the principled fight against polio was in large measure a function of his loyalty to FDR.

Morris Fishbein, MD (1889–1976), editor of the *Journal of the American Medical Association* and long-time advisor to Basil O'Connor, recognizing the difficulties of biographical portraiture, stated "Mr. O'Connor is a somewhat reticent man, … and he has never been given to reminiscing concerning his origins, his education, his personal life, or his accomplishments in the many fields in which he has toiled." Indeed, the parsimonious trail of O'Connor's personal papers leaves the biographer with scant traces of his early years and much of his personal life. Portions of extant collections of his papers have been dispersed through five repositories, and others known to have existed are unaccounted for and presumed lost.[2] According to the Rauner Special Collections Library of Dartmouth College, the estate of Basil O'Connor destroyed O'Connor's personal and family papers at the time it donated the existing collection of Basil O'Connor Papers (relating largely to FDR, Dartmouth, and the Red Cross) to the college in 1972.[3] The extant collections are not abundant in information about his personal life, which he protected carefully as "off limits" to the public at large, and he left behind a documentary record consisting largely of official memoranda and correspondence, speeches, and the corporate records of the NFIP, GWSF, and Red Cross.

---

[2] Archival repositories utilized in this study include the March of Dimes Archives (MDA; White Plains, NY), Franklin D. Roosevelt Presidential Library (FDRL; Hyde Park, NY), New York State Archives (NYSA; Albany, NY), Roosevelt Warm Springs Institute for Rehabilitation (RWSIR; Warm Springs, GA), the American Red Cross Archives (ARC; Lorton, VA), and the Rauner Special Collections Library of Dartmouth College (DART; Hanover, NH).

[3] Private communication, Rauner Special Collections Library; May 2015.

Morris Fishbein's characterization treats only the outline often repeated elsewhere, a standard formulation of O'Connor's rags-to-riches saga:

Basil O'Connor was born January 8, 1892 in Taunton, Massachusetts. He was the youngest of the five children of Daniel Basil and Elizabeth Ann (O'Gorman) O'Connor. His father was a tinsmith, never financially well off. Basil was actually christened for his father, Daniel Basil O'Connor, but early in his career he dropped the Daniel when he saw how many other Daniel O'Connors there were in the New York telephone book.

Basil had the typical childhood of any American boy born under such circumstances. He peddled newspapers to help support his family. As a student, [he] early showed the ability to grasp information quickly, to retain what he had learned, to be healthily skeptical of what was not well established. There is no record of his intelligence quotient, but he skipped at least two grades in school and was admited to Dartmouth College at the age of 16. For some reason he learned to play the violin. This seems to have been most fortunate, because payment for his playing aided his maintenance during his years in college.

At Hanover, New Hampshire, he is remembered as a slight, not prepossessing boy weighing only 111 pounds; a classmate called him "the skinniest freshman ever seen." The football coach at Dartmouth at that time was one Doc O'Connor, so the students promptly nicknamed Basil "Doc." The appellation has been retained since that time by most of his close friends. Nowadays, however, people think that he is called Doc because of his great leadership in research and education, not only in poliomyelitis but in medicine generally.[4]

Of his home life we know next to nothing; of his years at Dartmouth College and Harvard Law School, we know little more. His parents were both born in Ireland; his father Daniel Basil O'Connor, Sr. found employment in various iron and steel mills in Taunton. O'Connor once commented that his father earned basically the same wage his entire life, a static condition of earnings that his son quickly overcame once launched as an attorney. His family certainly knew the pinch of poverty; for lack of winter clothing, O'Connor's mother would stuff his coat with newspapers to provide a little extra warmth during the freezing walk to and from school. The young Basil developed his sharp business acumen in his high school years, becoming the business manager for the Taunton High School Journal and paperboy for the *Taunton Gazette*. He also worked as an odd job painter and a sales clerk at Colby's Clothing Store in Taunton. It was at Dartmouth College that O'Connor found a secure footing in the world as well as his lifetime home and family—a connection to a network of like-minded friends and colleagues that he cherished and that lasted into his senior years.

---

[4] Morris Fishbein, "A Biographical Note." *Cellular Biology Nucleic Acids and Viruses.* (New York: New York Academy of Sciences, 1957) p. 3.

Basil O'Connor with High School Band, 1908. Basil O'Connor, seated front row center holding violin, with members of the Taunton (MA) High School band. O'Connor's reputation as a musician lasted through his years at Dartmouth College, for which he formed a passionate attachment. As an alumnus O'Connor rarely missed a Dartmouth commencement or major football game within his reach, and his network of Dartmouth alumni formed a vital proving ground for the development of his fund-raising acumen. He said, "I like to make fans of other people."

At Dartmouth O'Connor was an honor student who won prizes in two debating societies and was a member of the Webster Club, Sigma Phi Epsilon, and Delta Sigma Rho.[5] Geoffrey Ward characterized him as "… a fast-talking tinsmith's son … who had entered Dartmouth at sixteen and earned the title 'Most Likely to Succeed' after putting himself through school by playing the violin in a dance band he'd organized himself; he had even launched his own fraternity after the established ones turned him down. When too much study at the Harvard Law School rendered him temporarily blind, he kept up with his classes by persuading fellow students to read to him."[6] O'Connor attended Dartmouth College from 1908 to 1912, graduating in seven semesters.

[5] *Dartmouth Alumni Magazine*, May 1972, p. 82.
[6] Ward, Geoffrey C. *A First-Class Temperament: The Emergence of Franklin Roosevelt.* (New York: Harper Perennial, 1989) p. 657.

His major was political science, with minors in Latin and biology. He tended to favor languages early in his undergraduate career, taking courses in Greek, English, and German in addition to Latin. His grades were unremarkable: an "average standing" (grade point average) of 78 upon graduation, and there is nothing in his college transcript that heralds the leader and celebrity he was to become. He graduated 65th in a class of 222; he chalked up quite a few absences from class, several due to illness; and he played violin in the college orchestra in all but his last semester.

The simple facts recorded in his transcript reveal nothing of O'Connor's profound connection to the spirit of Dartmouth College and the enduring friendships with his classmates. Basil O'Connor became a Dartmouth booster *par excellence* and his active concern for the welfare of the college as an enthusiastic alumnus in the decades to follow ties him intrinsically to the history of the college, its alumni associations, and its library and archives. His undergraduate education was a formative social experience with many friendships, and after 1912, the year of his graduation, he gravitated deeper into a matrix of rich personal associations in a far-flung network of alumni. As President of the Class of 1912, his connection to fellow alumni continued unabated through constant letter-writing. Even during his years at Harvard Law School and immediately thereafter as he began to carve out a career for himself in the legal profession, the volume of his Dartmouth correspondence was prodigious. In personal letters and even in fund-raising appeals he was informal and playful, and he readily accepted challenging responsibilities, such as the charge of raising $25,000 for the 25th class reunion in 1937. In 1917, he was the Class Agent; in 1918, he headed the Dartmouth War Fund after the United States entered World War I. O'Connor became legendary as Dartmouth's "Delegate-at-Large from Hanover" and the "Number One Dartmouth Fan." He became such a general spokesperson in his many alumni roles that one friend styled him the "super-orator, publicity chairman, the one and only living specimen, Doc O'Connor." A truly devoted fan, O'Connor attended every major Dartmouth football game, often traveling long distances with an entourage of family and friends, and missed only one annual commencement ceremony in the 30 years leading up to his installment as president of the NFIP.[7] He could always be relied on to support Dartmouth. The very month that the NFIP was founded, January 1938, he donated $500 to help launch the Friends of the Dartmouth Library.[8]

---

[7] DART, Basil O'Connor Papers, General Correspondence, folders 1 to 4.

[8] West, Herbert F. *The Impecunious Amateur Looks Back: The Autobiography of a Bookman*. Hanover (NH): Westholm Publications, 1966, p. 151.

The roster of O'Connor's formal positions as a Dartmouth alumnus is formidable. During the years of the Roosevelt and O'Connor law partnership and involvement in the GWSF with FDR, he was a member of the Dartmouth Alumni Council (1923–29), President of the Dartmouth Alumni Association of New York (1929–32), and President of the Dartmouth College Club of New York (1928–32). In later years he was closely involved as a member of the Executive Committee of both the General Association of Alumni and the Dartmouth Medical School Campaign (to launch the medical school), and in 1938 led an alumni committee for construction of a new theater and auditorium. Dartmouth awarded him an Honorary LLD (Doctor of Laws) degree in 1946, he received the prestigious Dartmouth Alumni Award in 1957, and as President of the NFIP he returned in 1962 on the 50th reunion of the Class of 1912 to deliver an address on the diminution of individual liberty in a world of technological change titled "The Diminishing Citizen." He was driven to forge lasting personal connections and to nurture them continuously. It is easy to believe that if O'Connor was alive today, he would be completely in synch with the personal networking potential of email and Facebook. O'Connor utilized his powerful network for fund-raising for the GWSF and NFIP; many of his classmates faithfully contributed to the March of Dimes. He assiduously pursued the hobby of collecting photo portraits of Dartmouth alumni, never embarrassed to make repeated requests to build his personal portrait gallery. His collection of Dartmouthiana—prints, books, photographs, manuscripts, and publications—was considered the largest outside of Hanover, New Hampshire, and much of it returned to the Rauner Special Collections Library after his death in 1972.

If Dartmouth prepared Basil O'Connor for the world, Harvard prepared him for a career. O'Connor graduated from Harvard Law School in 1915, and his subsequent ties to Harvard were not quite as profound as with Dartmouth, though he later became a trustee of the Harvard Law School Association of New York City. Upon graduation from Dartmouth, O'Connor and his three close friends Conrad Snow, Leslie Snow, and Henry Urion (who later joined O'Connor & Farber as an attorney in 1935), in a moment of ebullience, wagered on which of the four would first marry, the stakes being "drink and dinner, the same to be paid for from his own exchequer, whenever he shall enter into lawful wedlock."[9] Whether he lost the bet is unknown, but O'Connor married Elvira R. Miller of Louisville, Kentucky on August 31, 1918; the couple had two daughters, Bettyann and Sheelagh.

[9] RWSIR, Box 33, Folder 17. July 4, 1912.

He was admitted to both the Massachusetts bar (1915) and New York bar (1916), first practicing law with Cravath and Henderson in New York, and returning to Boston in 1916 to work with the firm of Thomas Streeter and Robert Holmes. His law practice focused on arranging contracts between oil companies and refiners; and he made regular trips to Oklahoma to certify legal qualifications for corporations doing business in the state. With this expertise he later advised FDR on a host of legal matters ranging from business contracts to foreign oil concessions. The *Dartmouth Alumni Magazine* characterized him in 1938 as "one of the best known lawyers on oil matters around town."[10]

Thomas Winthrop Streeter (1853–1965), a lawyer and banker, had an inestimable impact on Basil O'Connor's life. The two first met in the aftermath of a debating competition at Dartmouth in which Streeter, as an appointed judge, had decided against him. O'Connor's loss sorely disappointed the Dartmouth debating team and set the entire student body into an uproar. Taking O'Connor into his confidence afterward, Streeter admitted in a private meeting with the young debater that an "injustice" had occurred. So impressed was he with O'Connor's intelligence, drive, and vision that Streeter ended up helping to finance his Harvard education toward his law degree. His employment at Streeter and Holmes was a consequence of that fortuitous encounter, and O'Connor parted company with Streeter only to open his own law firm in New York in 1919. The two remained lifelong friends, and when Streeter and his wife lost their entire fortunes in the stock market crash of 1929, O'Connor promptly gave him $50,000 from which he rebuilt his fortune.[11] Later, in 1943, O'Connor repaid him for his original generosity once again by recommending Streeter's wife for the post of Director of the Marine Corps Women's Reserve. Settled in New York City, O'Connor practiced law under his own name until forming the partnership Roosevelt and O'Connor in 1924. O'Connor's original gift to Thomas Streeter came full circle in 1965 upon his death in a $50,000 bequest from Streeter to Dartmouth College "to establish a scholarship fund in honor of my friend Basil O'Connor of the Class of 1912."[12]

Streeter's support arrived at a critical point in the transition from academia to law career, and their lifelong friendship endured. O'Connor's friendship with FDR was also momentous and life changing. Like O'Connor, Roosevelt could boast hundreds of friends, and it is instructive

---

[10] *Dartmouth Alumni Magazine.* December 1938.
[11] MDA, Oral History Records. Hazel Dillmeier interview. February 28, 1986.
[12] DART, Basil O'Connor Papers, General correspondence.

to dwell a moment on the character of friendship in the highest echelons of political life at that time. Another Harvard Law School graduate of that age, the jurist Felix Frankfurter, whom FDR appointed to the US Supreme Court in 1939, provided an insightful appraisal of his own friendship with FDR: "Franklin Roosevelt was very active and full of ideas, full of interest, and we became friends, close friends, the way men with genial common factors become friends when engaged in a common enterprise, and those active years of comradeship in important work of an official kind rather transformed a casual, pleasant relationship into what might be called a warm friendship."[13] This sort of "power friendship" in the matrix of American politics was embedded in upper-class relations that unified the structure of wealth, education, and business, and that sociologist E. Digby Baltzell described so trenchantly in his classic account *The Protestant Establishment: Aristocracy and Caste in America*. If the enclosed structure of the Protestant establishment, to which FDR belonged by virtue of his familial, Groton, and Harvard networks, was generally challenged by a rising meritocracy to mitigate its inbred nature and infuse it with democratic tendencies, we might single out the Roosevelt–O'Connor friendship as an ideal example of how this played out between highly placed individuals. Their projects of formal partnership—their law firm, the Brain Trust, Warm Springs, and the March of Dimes—served to cement their friendship within the ramified network of many other personal, social, and business connections.

The chronological course of his relationship to FDR is recounted in the chapters to follow. O'Connor's personal character, management style, and worldview deserve an examination on their own merits, though often difficult to isolate entirely from the influence of Roosevelt. O'Connor was most well known for his leadership of the NFIP from 1938 to 1972 and the triumph of the Salk vaccine in 1955; it is a cliché to speak of him as the "architect in the fight against polio" or the "arch foe of infantile paralysis": these facile characterizations have been overused. In his maturity, his personal character was defined by his shrewd intellect, obsessive devotion to work and study, and upward mobility from his working-class origins. He cultivated the appearance of success in compensation for his humble origins, but also because he grew increasingly status proud through the phenomenal success of his legal work. His urge to take up humanitarian causes developed as his friendship with FDR matured, and though he espoused a strong belief in participatory democracy, he refused to abandon

---

[13] Phillips, Harlan B. (editor). *Felix Frankfurter Reminisces: An Intimate Portrait as Recorded in Talks with Dr. Harlan B. Phillips*. New York: Doubleday, 1962. pp. 277–78.

a rigid hierarchical model of organizational leadership, which he applied effectively as head of the NFIP though not quite as successfully at the helm of the Red Cross. O'Connor's admiration of FDR's experimental approach to the socioeconomic catastrophe of the Great Depression through New Deal programs leavened his authoritarian mind set with respect for intellectual expertise and the free gift of volunteerism: key ingredients to his concept of meritocracy in which he reconfigured volunteerism and the role of the informed layman as a type of elitism that tended toward, but did not quite equal, the discipline of intellectual accomplishment. This was a key philosophical underpinning of the March of Dimes as a social phenomenon. Though his success as an attorney seems overshadowed by his public achievement of the March of Dimes, both were rooted in an indomitable work ethic. In a biographical portrait in 1941, the *Sigma Phi Epsilon Journal* characterized O'Connor as the "easily the hardest working member of either staff," referring to his law practice (O'Connor & Farber) and the NFIP.

In her history of the Salk Institute for Biological Studies, Suzanne Bourgeois characterized Basil O'Connor as a "venture philanthropist." O'Connor's launch of the NFIP is illustrative in this respect. He quickly assembled a board of trustees in late 1937 cultivated from his partnership with FDR, personally handled the incorporation of the foundation, and within 18 months had populated the NFIP advisory committees with a veritable "*Who's Who*" of leading experts in medicine and science. He led the difficult project of merging a loose infrastructure of volunteers loyal to the President's Birthday Ball fundraiser into the NFIP structure of county chapters. During the war years and afterward O'Connor crisscrossed the nation by train to "gather the power of volunteers." In the Red Cross years (1944–49) his focus was diversified and his outlook was more international; but the years leading up to the Salk vaccine were devoted to national travel, months in advance of the annual March of Dimes every January. His speech-making to community leaders and potential volunteers had all the trappings of a political campaign. He patiently explained the years of painstaking work with FDR building the polio center at Warm Springs, cogently emphasizing the "full realization of the tremendous problems that constantly confront those afflicted with infantile paralysis."[14] While outlining his plan to stop polio O'Connor went even further to convey the Rooseveltian ideal of "freedom from disease" through the NFIP mission which he believed might also bolster the civic structures in communities

---

[14] MDA, Basil O'Connor Papers. Series 1: Typescripts. *The National Foundation for Infantile Paralysis.* April 10, 1939.

throughout the country toward the end of "mastery of other diseases as well."[15] Thus, future expansion of the NFIP mission beyond polio was part of O'Connor's outlook—and to his way of thinking, completely within reach—from the very beginning, since the NFIP, born midway through Roosevelt's 12 years as president, had all the hallmarks of a national movement congruent with the general purpose of the New Deal. While many saw the NFIP simply as "the polio foundation," O'Connor promoted it as a national civic organization.[16]

O'Connor took to the stump politicking for polio prevention in March of Dimes campaign tours and yet he would dilate his focus to address the organic development of volunteers in communities, carefully suggesting the possibility of a national program based on a "spontaneous mass endorsement" necessary for public health, and not just a "city council, county board of supervisors, or state legislature to appropriate tax funds for the support of a health department."[17] He exhorted his audiences to think above and beyond "charity" to promote active participation in the March of Dimes, which he came to believe was a key to participatory democracy. He never failed to personalize the message that the NFIP was "*your* National Foundation." The fundamental pitch was not merely to donate a dime, but to "*Join* the March of Dimes" for personal membership was the crux of further involvement. He delivered lectures to medical leaders, insurance companies, bar associations, religious organizations, and civic groups, touting the March of Dimes as "the only national voluntary agency fighting against a specific disease that pays the cost of patient care where necessary."[18] As potential volunteers coalesced into committed chapter representatives, his visits to chapters, from major market areas to the rural backcountry, continued to have tremendous value. In the 1950s, his encouragement of NFIP state representatives became particularly important, as the Salk vaccine loomed into view as the answer to polio. After the war and Roosevelt's death, parlaying his personal experience of seeing war's devastation in Europe and the Pacific, he was also wont to broaden his speeches to philosophize on the brotherhood of humanity, which he proudly claimed was the ultimate goal of his dear friend Franklin.

Charles Massey, who headed the March of Dimes from 1978 to 1989, commented that Basil O'Connor "always made a speech but was not a good

---

[15] MDA, Basil O'Connor Papers. Series 1: Typescripts. *Role of The National Foundation in a Public Health Program.* 1941.

[16] Rogers, Naomi. *Polio Wars: Sister Kenny and the Golden Age of American Medicine.* New York: Oxford University Press, 2014, p. 21.

[17] MDA, Basil O'Connor Papers, ibid. 1941.

[18] Ibid., Speeches for 1952 March of Dimes. 1952.

speaker," and Dorothy Ducas, Public Relations Director, noted that
O'Connor, while extremely adept in selling the March of Dimes, was actually rather inept in "selling himself." His March of Dimes and Red Cross
speeches were organized like legal briefs and delivered with businesslike
aplomb but little vivid emotion. Doc's lack of rhetorical flamboyance in
speech-making in preference for sticking to "the facts" was contraindicated
by his everyday sartorial elegance. The fresh white carnation, inevitable cig-
arette holder, tailor-made clothing, membership in the Sky Club and Fair-
way Yacht Club, and solid preference for the Waldorf–Astoria as a standard
venue for any social or business occasion all advertised his status-proud
narcissism. Massey recalled his first meeting with Basil O'Connor, receiving
him as the Arkansas State Representative in 1949:

> Prior to O'Connor's visit I received a two-page briefing sheet from his office secretary,
> Maggie Eagan, and I will never forget the details. For example, there should be an
> Underwood typewriter (not a Remington) in his suite, along with a supply of Her-
> shey chocolate bars. He was on and off the wagon in those days so there should be
> bottles of ginger beer and a bottle of Cutty Sark scotch, just in case. And, most
> important, the New York Times should be delivered to his suite each morning,
> along with a white carnation for his lapel – a fresh carnation, with a natural stem,
> no wires.[19]

The appurtenances of wealth and even dandyism stand out in high relief
in the popular memory of Basil O'Connor, conjoined to the ready-made
stereotypes of Doc as FDR's law partner and as the nemesis of poliomyelitis.
Naomi Rogers notes that "[w]hile his dapper clothes and luxurious life-style
were at times attacked as signs of 'excessive fund raising and egregious razzle-
dazzle,' O'Connor boasted that he received no salary from the NFIP and that
his insistence 'on traveling de luxe' was a deliberate tactic designed to show
everyone that the NFIP was 'something special.'"[20] Dorothy Horstmann of
the Yale University Poliomyelitis Study Unit wondered with amazement
that O'Connor had hung a large, conspicuous portrait of himself opposite
his office desk, and Massey commented that O'Connor's reputation as a
"vain peacock" easily compared with General Douglas MacArthur. Though
his mastery of law, the news media, and collaborative politics paved the way
to very public forms of success, Massey reflected that on first impression it
seemed to many people "discordant that a person with his flamboyant style
was the head of an organization raising nickels and dimes to fight polio." For
this reason Dorothy Ducas believed that he always required "interpretation."

[19] Massey, Charles. Personal interview. July 22, 2004.
[20] Rogers, Naomi. ibid, pp. 31–32, n. 97.

O'Connor's underlying humanitarianism and responsibility for the practical successes of the NFIP had acquired this paradoxical aspect:

> I think he saw the [chapter] visits as a duty, [ie] "noblesse oblige." I don't believe he listened to staff members or took their pulse. I say that because he never invited questions. The protocol at field staff and volunteer meetings included a receiving line so everyone could line up and shake hands with the president. Usually there was a dinner at which he was the featured speaker. He was treated almost like a monarch. … There is little doubt that O'Connor deliberately cultivated the imperial aura that later worked against him. At that time, however, he was presented as a successful, wealthy Wall Street lawyer in a personal crusade against polio. He was the supreme commander of an army of volunteers.[21]

O'Connor constantly sought the counsel of experts in every subject that touched on polio and the NFIP big picture but placed scant value on advice from his own staff. He sought opinions and informed advice from ranking staff members, but decision making was strictly top down, never by consensus. This style of leadership caused an unrelieved undercurrent of dynamic tension, even though many ranking executives and directors—Charles Massey, Joseph Nee, Dorothy Ducas, and Elaine Whitelaw—shared O'Connor's own strong belief in the Foundation and unshakable commitment to stop polio. Staffers would ask "Who has Doc's ear this week?" or "Who's in the doghouse?" Massey characterized his style of leadership as exceedingly lonely:

> O'Connor's style was to be in total charge with everyone else in orbit around him. Even his top administrative assistant was in the orbit, whether it was Joe Savage when I joined the Foundation in 1948, Ray Barrows in the early 1950s, or Joe Nee in the 1960s. None of them was allowed to be a real manager let alone a leader. They were little more than assistants, charged with relaying O'Connor's instructions and carrying out his orders. In short, theirs was a difficult, if not impossible, task, with responsibilities but no real authority.[22]

Massey's executive perspective derives from years as O'Connor's right hand after the mission change to birth defects in 1958, a perspective he shared with Joseph Nee and William R. Russell as they tried carefully to steer O'Connor's increasingly sclerotic opinions about government and management through moments of controversy during a difficult period of organizational change. Massey had worked as NFIP Regional Director and Assistant Director of Chapters in the 1950s, but when the change in mission direction occurred, he appreciated the independence of his "orbit around" O'Connor:

> I was fortunate during those years because O'Connor was unaware of what I was doing and he left me alone. Consequently, I was able to initiate a number of projects

---

[21] Charles Massey, ibid.
[22] Ibid.

that he would not have dreamed of: collecting surplus funds from chapters to meet the budget; restructuring our field organization; and consolidating smaller chapters into major market chapters. These and other initiatives were taken soon after our new mission was announced in 1958 when it became clear that we could no longer do business as usual. While such initiatives were vital to the foundation's survival they were of little interest to O'Connor. Having announced a long-range plan he just expected it to be implemented. For me, his disinterest in the process was a blessing.[23]

Massey's perspective is but one of several notable staff perspectives on Basil O'Connor. Many felt drawn by O'Connor's personal magnetism and dedication (as Massey himself was) despite his obvious hauteur. Most believed that he above anyone else in the organization imparted a profound sense of ownership in the polio mission. Though intolerant of informality and very critical of error, his sense of humor and camaraderie would shine through at sudden moments during off-hours or on social occasions. His impeccable dress, FDR connection, and grandiose style of presentation led to a veritable folklore of stories about Doc O'Connor among NFIP staff and volunteers. Socially well connected, his Christmas card mailing list boasted 2,500 recipients. He delighted in talking at length about his friendship with FDR, and when asked about the fact that he was FDR's former law partner, he would always insist on the opposite: "No, FDR was *my* law partner!" His idiosyncrasies of managerial style, such as the private intercom he used to communicate with staff, known with general dismay as "O'Connor's squawk box," became a workplace embarrassment. When one of his underlings had the temerity to ask him to dictate a recently given speech in his "spare" time, O'Connor replied testily, "We great men have our moments—and if you don't record us in them, *then you don't record us in them!*"[24] Was this megalomania or mere peevishness? Fellow attorney and NFIP general counsel Stephen V. Ryan would answer that neither of these applied. Ryan characterized Basil O'Connor as utterly fearless and perpetually curious, that his ability to penetrate the extraneous verbiage in any discussion to get to the gist of an issue was itself legendary. Ryan considered him "one of the smartest lawyers I ever knew," even though the opinion might be divided whether he was a gentleman or SOB. Gabriel Stickle, NFIP Statistician who eventually became O'Connor's Executive Assistant, described his first encounter with O'Connor that was probably quite typical: he answered his telephone one afternoon in 1951 to hear a gruff voice ask, "Do you know anything about statistics?" Instantly recognizing his

---

[23] Charles Massey, ibid.
[24] MDA. Series 7: Memoranda. April 7, 1947.

voice Stickle replied, "What would you like to know, Mr. O'Connor?" "What's a geometric mean?" Stickle continued: "I hesitated for a moment and then gave him a one-sentence definition and a three-number illustration. There was a grunt on the other end of the line, and he hung up. He was running into the concept of geometric mean antibody titer, antibody titer measured in terms of the geometric mean titer." Stickle continued:

> … shortly thereafter he contacted the biostatistician at Johns Hopkins and that person gave Mr. O'Connor a textbook on statistical methods, and he soon began to spout statistical terminology such as "universe," meaning the population from which a sample is drawn. And there is on my wall even now a photograph of Basil O'Connor with the greeting, "From one universe to another, but both of them rather good; signed, Basil O'Connor."[25]

Beneath the overt contradiction of O'Connor's autocratic, paternalistic style of leadership vis-à-vis his professed humanitarianism lay the broad-minded progressivism of Roosevelt and his polio disability. Much has been written detailing FDR's personal recovery from polio, as well as its effects on his character and political career, but it was Roosevelt and O'Connor *together* that collaborated most closely for the *nation's* recovery from the disease. O'Connor's own personal sympathy for Roosevelt's disability is carefully sublimated in his public esteem for FDR's capabilities. His praise and devotion included an appreciation of FDR's visionary capability to perceive the strategic means to bring an end to polio that unfolded gradually over the course of three decades. The aftereffects of polio also brought the plight of disability both as individual tragedy and a public spectacle into everyday view. As we will see below, FDR's project to make America "polio conscious" helped to bring awareness of the disabled as a minority group with special needs and rights into even sharper focus. Consequently, as FDR dragged the nation up from the morass of economic collapse, there was a gradual loosening up in O'Connor's paternalism toward other underrepresented groups, particularly African-Americans and women. This was driven, in part, by the legitimate criticism of Warm Springs as a segregated facility. His brief stint with the National Conference of Christians and Jews in 1940, at the very moment when the racist totalitarianism of the Nazis seemed poised to engulf the world, was a pivotal experience in his public embrace of principles of equality and fairness. By the end of the war, and with Roosevelt's death, his public pronouncements on race and class began to have all the rhetorical hallmarks of the developing egalitarianism of the

---

[25] MDA, Oral History Records. Gabriel Stickle interview. August 17, 1983.

burgeoning civil rights movement. In an editorial for *The New York Age* in 1946, O'Connor wrote, "The National Foundation is one of many groups in this nation that have discarded the ignoble fetish of race, religious or class distinction, are devoted to serving human beings—crippled, blind, sick, undernourished, illiterate, or poor."[26]

In his pronouncements on tolerance and race, which continued during this leadership of the NFIP and association with the Tuskegee Institute as board chairman, he said little about women as a minority or the question of women's rights. In her biography of Sister Elizabeth Kenny, Naomi Rogers recognized O'Connor as a "tough-minded Irish Catholic in an era when Catholics faced widespread social discrimination."[27] Indeed, the young Basil O'Connor had felt the sting of such discrimination and inequality in the bare circumstances of his family's struggle to make ends meet, and it is instructive to consider how his heightened consciousness of inequality at mid-life shaped his relations with women. Though married with two daughters, O'Connor spent little quality time in the domestic setting; he was a known workaholic before that term was invented, and he was always on the move. He thrived in a man's world, and the legal "fraternity" and Dartmouth College had both excluded women, despite the gains of the suffrage movement (Dartmouth opened admission to women in 1972). On the other hand, he had much to observe in FDR's example. Roosevelt was bound by very strong familial ties to his mother, delighted in the social company of women throughout his life, and made strides to include women in his administration. One thinks of Roosevelt's marriage with Eleanor Roosevelt, whose spirit of independence and social activism was reflected in exceedingly complex ways through, over, and against FDR himself. FDR's secretarial staff became a familial cadre in the White House: Marguerite (Missy) LeHand had assisted FDR in the most intimate ways from the time he was stricken with polio. As a consequence Missy LeHand has been called one of the most powerful women of the 20th century in her critical support of FDR though largely unappreciated for that. FDR's long-standing relationship with the ever-dependable Frances Perkins led to her selection as the first female cabinet member in US history; she was Secretary of Labor for the duration of the Roosevelt presidency. Distant cousin Daisy Suckley became FDR's "closest companion," a reliable family confidante to the bitter end. Though O'Connor's own relationship with

---

[26] DART, Basil O'Connor Papers. "Building Blocks for Tolerance," editorial, *The New York Age*. September 28, 1946.
[27] Rogers, Naomi. ibid, p. 16.

Eleanor Roosevelt, as we shall see, was at first skeptical and testy, it eventually matured into a lasting friendship. His personal validation and support of women as equal contributors in work, opportunity, and professional success reached a climax in his personnel choices at the NFIP. This was a little recognized but very notable achievement, for just as the Dartmouth connection was predominantly male oriented, so too was the NFIP board of trustees for the first three decades of its existence.

Basil O'Connor and Daughters, c.1936. Basil O'Connor was a family man and an enthusiastic fan of Dartmouth College football. Here he enjoys a football program with his daughters Sheelagh (left) and Bettyann (right). Bettyann was later stricken with polio in 1950, to recover at Warm Springs under the care of Dr. Robert Bennett. O'Connor's lavish method of supporting Dartmouth football often involved renting a private railway car with an invitation to 50 or 100 friends to travel to Princeton, Ithaca, or Boston to see "the Big Green" team on the gridiron.

If the Roosevelts together raised O'Connor's consciousness about opportunities for women in the 1930s when women suffered disproportionately from discrimination in employment when there were few jobs to be had anywhere, he also benefitted from the successful example of women like Alice Plastridge, who was the paragon of physical therapy administration at Warm Springs. When the time came to create the NFIP medical committees in January 1938, O'Connor queried his advisor Morris Fishbein about the selection of qualified women. Fishbein equivocated, "I would hesitate to mention the names of women physicians who might be appointed on some of the various committees of the National Foundation although I think the idea in general is a good one." He recommended that only the chairs of the Committee on Nutrition and the Committee on Epidemiology find women qualified to serve, doubting that any qualified female would be found in the fields of orthopedic surgery, virology, and medical publications, which was Fishbein's own specialty as editor of the *JAMA*.[28] Fishbein's lackluster endorsement expressed just the sort of skepticism about qualifications and credentials that tended to protect paternalistic ways of thinking, which in turn protected the structure of job discrimination against females. This was just the sort of barrier that Virginia Apgar, creator of the APGAR score, experienced when she was advised to turn away from surgery as a medical specialty in the 1930s and that eventually led to her career as an obstetrical anesthesiologist (she later became Vice President of Medical Affairs at the March of Dimes).

If O'Connor failed to populate his original board of trustees and medical advisory committees with women at the founding of the NFIP, his ideas about social justice began to find a new determination with the advent of the war and the Nazi menace. His leadership to overcome the legacy of segregation at Warm Springs led to his permanent involvement at the Tuskegee Institute on the heels of a major NFIP grant to build an infantile paralysis center there, and his work with the National Conference of Christians and Jews explicitly promoted the ideology of tolerance and social justice at a very dark time. Moreover, he found ample opportunities to promote minorities in the ranks of the NFIP itself. He hired the African-American civil rights leader Charles Hudson Bynum to fill the post of Director of Inter-racial Relations to uphold the NFIP pledge that polio care would be given without regard to "race, color, age, or religion." This important decision ensured that polio care was positioned as a civil rights issue. O'Connor also delegated some of the most important responsibilities of the new foundation to women. Dorothy Ducas, a reporter for the *New York Herald Tribune* in the 1930s, became the NFIP Public Relations Director from 1946 to 1962, one of whose later contributions was to coin the term "birth defects." Catherine Worthingham, president of the American Physical Therapy Association and Professor of Physical Therapy at Stanford University became Director of Professional Education from 1944 to 1961. Worthingham's program of physical and occupational therapy education through NFIP scholarships, fellowships, and grants effectively trained over

---

28 MDA, Medical Program Records. Series 5: Committees. Medical Advisory Committees, 1938–1959. "Women on Medical Committees." September 15, 1941.

2,600 physical therapists in the 1950s, one-third of the active physical therapists in the nation at the time. Most of these were women. Elaine Whitelaw (1914–92) was a nationally recognized leader of volunteers whose NFIP career spanned five decades, from 1943 to 1990. As Director of Women's Activities, Whitelaw befriended and nurtured women from all walks of life to take up volunteer leadership roles in communities across the nation. These women became vitally important in the fight against polio as fundraisers and as leaders of Polio Emergency Volunteers (PEVs) battling polio epidemics at the community level. Elaine Whitelaw appealed to women by combining the ethics of community service, the glamor of fashion shows, and the humanitarianism of science in service to people. A case could be made that Whitelaw's empowerment of female volunteers in the March of Dimes in part contributed to the wave of feminist activism that emerged in the 1960s after the "Rosie the Riveter" gains of the war years had lapsed into male chauvinist complacency nested under the façade of domestic tranquility. But Whitelaw would never have allowed any perception of her March of Dimes role as that of a "feminist." According to Charles Massey, she wanted to be "one of the boys."

Elaine Whitelaw's account of her first encounter with "Roosevelt's law partner" opens a tiny portal into the brusque but calculating personality of Doc O'Connor. The scene was O'Connor's office in the Equitable Building at 120 Broadway in Manhattan, the site of the Roosevelt & O'Connor law office that became national headquarters of the NFIP. Completed in 1915 and now a National Historic Landmark, the Equitable Building was then the world's largest office building by floor area. In 1943, Whitelaw had been working on a temporary assignment for the NFIP and was thrilled with the promise of the new foundation's fight against polio. She wanted to do more and insisted to Peter A. J. Cusack, Executive Secretary of the NFIP, on meeting the President of the Foundation to discuss her ideas for creating a cadre of women volunteers:

> I was enthralled because the first thing I saw was Franklin Roosevelt's office which had been left untouched from the day that he went to Albany as Governor of New York. There it was, and it was very exciting and moving. Well, we came into the office and sitting behind the desk was this dapper, immaculate, shining, then fat, man. Peter Cusack said, 'Mr. O'Connor, this is Ms. Whitelaw.' He never rose, he didn't say how do you do, hello, or nice to see you won't you sit down. So, I just stood there. Peter sat down [and] so did I in front of the desk. The first word [O'Connor uttered] was, 'Well?' I have never seen anything like it: the lack of courtesy, the lack of welcome. So I began to talk and say what I thought about, what I could be doing with all these women volunteers. Then suddenly, I looked up at the clock, and I realized I had been talking

for 14 or 15 minutes. I stopped abruptly, and I leaned across the desk and said, 'Well?' And I honestly believe it was from that moment that a long and wonderful friendship began. He was so taken aback, he began to sputter, and he said, 'That's all very interesting, I'm sure it's a lot of boloney, but it's interesting, and let's talk about it.' We talked for three and a half hours. He was really marvelous.[29]

Whitelaw was totally overcome with the peculiar intensity of the encounter. After the meeting, she staggered out of the office toward the elevator holding onto the marble walls of the hallway for support and waited for Peter Cusack to finish a private consultation with O'Connor. Cusack treated her to a drink afterward and told her, "All Mr. O'Connor could say is 'don't let that girl go—she will fight for what she believes.'" Whitelaw went on to an illustrious March of Dimes career of nearly half a century, fighting for what she believed. Perhaps the first occasion of a woman literally "leaning in" could be dated from her interaction with Basil O'Connor.

From the first years of his association with FDR at Warm Springs to the period when he guided the transformation of the NFIP to birth defects prevention and helped to create the Salk Institute for Biological Studies, Basil O'Connor developed a strategic vision of how to identify and counteract the determinants of disease. Dr. Edward Tatum of the Rockefeller Institute went so far as to claim that "O'Connor practically created virology … not by concentrating research grants narrowly on polio, but by encouraging the most basic studies of all viruses."[30] His ability to telescope this vision from the widest perspectives to the minutest details sometimes led to bold strokes of policy. His decision in 1954 to direct the emerging pharmaceutical companies to produce enough polio vaccine *before* the Salk vaccine proved effective seemed a daring gamble at the time. He justified this decision as absolutely necessary for those children who failed to benefit from a true vaccine injection in the field trial. The authorization by the NFIP board of trustees stated:

… to negotiate contracts with the various drug houses for the preparation of a stockpile of poliomyelitis vaccine sufficient for use in 1955 among those children who participated in the vaccine field trial but did not receive the vaccine, all children in the first grade of school, and among pregnant women; that there be, and there hereby is, appropriated a sum not to exceed $10,000,000.00 for the purchase of such vaccine …[31]

O'Connor, fixated on his obligations and responsibilities as the appointed representative of the American people through the March of Dimes, was

[29] MDA, Oral History Records. Elaine Whitelaw interview. June 8, 1984.
[30] MDA, Basil O'Connor Papers. Series 8: Biographical Data. "One Man's War against Disease," *Medical World News*, 5(3). January 31, 1964.
[31] MDA, Board of Trustees Records. No. 6. June 18, 1954.

naturally thinking of children and women first. This was not immediately apparent for those who preferred to remain fascinated with his exterior—ie, the impeccable three-piece suit or tuxedo graced with a fresh white carnation—but an empathic charitable impulse was his raison d'etre and motivating force for all of his volunteer projects outside the practice of law.

In his final years, his strategic vision failed him, or rather, it failed to keep up with the times. He began to self-publish his speeches and mail them out as pamphlets to friends and colleagues. He spoke of "the heightened requirements of leadership," and he railed against conformity, domination by bureaucracy, apathy, specialization, and rapid technological change. He advised specific antidotes to these dire trends and always encouraged a broad, holistic view of social and political life. He championed the role of the "layman" in the advancement of science, claiming that "science is too important to be left to the scientists."[32] This opinion was not prone to endear him to certain scientists, but he had always characterized himself as a layman in the face of scientific complexity. He lashed out at what he called "government by elite vigilantes" and during his waning years grew further out of touch with March of Dimes executive staff who, as Charles Massey observed, took up the torch of everyday management virtually behind O'Connor's back both to protect him and to keep the organization effective and engaged. Basil O'Connor, a creature of politics himself and one of the favored sons of the New Deal, remained opposed to creating an advocacy office in Washington, DC for fear of the Foundation being seen as too "political." In this, he sensed that the encroachments of the federal government were poised to usurp the independence of organizations like the March of Dimes. And yet O'Connor felt free to leaven his stodgy diatribes and latter-day philosophizing with a twinkle of self-deprecating humor. At the Conference on Cellular Biology, Nucleic Acids and Viruses in 1957, he opened his address with an almost cosmic realization: "I, like the rest of you, am an accumulation of cells, an eye-dropper of nucleic acids, a few trace elements … and, I nearly forgot: probably some orphan viruses." On another occasion, before the United Community Funds and Councils of America, to whose philosophy of federated fund-raising he remained adamantly opposed, he candidly admitted his oppositional stance to his audience by saying that he was glad to be there "to let you see just how long my horns are."[33]

---

[32] MDA, Basil O'Connor Papers. Series 8: Biographical Data. Address, 65th Birthday Dinner. January 8, 1957.

[33] MDA, Basil O'Connor Papers. Series 3: Printed Literature. *The Place of the National Voluntary Health Organization in American Life.* March 5, 1959.

In 1953, Louis Finkelstein, Chancellor of the Jewish Theological Seminary of America, published a book titled *Thirteen Americans: Their Spiritual Autobiographies*. It included lectures given at the seminary's Institute for Religious and Social Studies by scholars and activists from all walks of life, presenting social and intellectual leaders "contributing to the preservation and advancement of civilization." Contributors included Clarence E. Pickett of the American Friends Service Committee, the astronomer Henry Norris Russell, civil rights activist Channing Tobias, and the Jewish scholar Judith Berlin Lieberman among others whose names might not be remembered today. It also included the reflections of Basil O'Connor, who admitted right off that the phrase "spiritual autobiography" terrified him. O'Connor's article, potentially full of promise for the biographer without access to the man's private thoughts and personal writings, consists of little more than a standard programmatic review of American values as might be presented in a high school civics class or college course on American democracy as understood in the 1950s. The "brotherhood of man," democracy as meritocracy, the cliché that work is the "law of happiness," and the necessity of fairness in human relations are all given a perfunctory, formulaic approval. O'Connor side-stepped any opportunity for genuine self-reflection; he revealed nothing of his inner life; he alluded to no spiritual quest or moment of doubt; he allowed not one shred of introspection. One is reminded of FDR's answer to a reporter's request to spell out his personal philosophy, to which FDR replied, "I'm a Democrat and a Christian—that's all." This "spiritual autobiography," if we may call it that, consisted only of a reiteration of core patriotic values derived from the American Revolution grafted securely to the Protestant work ethic. However, if we would dare to believe this single formulation is a complete summing up of his philosophy of life we would remain in error. His concluding thoughts in this brief moment of autobiography were these:

> Whether one has had a definite philosophy of life which one has followed unswervingly and continuously, or whether one has subconsciously pursued a line of activity that has imprinted its mark on one's existence, I cannot say. For myself I can say only that, because they were created and endowed by their Creator, I have always believed that all men are equal and are endowed with inalienable rights, including life, liberty, and the pursuit of happiness, and that those activities which have interested me most have been of a nature through which I believe men will be more likely to reach those ends.[34]

---

[34] Finkelstein, Louis., ed. *Thirteen Americans: Their Spiritual Autobiographies*. New York: The Institute for Religious and Social Studies, 1953. p. 229.

Throughout his life Basil O'Connor sought not philosophical depth, but action in an organizational context; not argumentation or endless preparation, but essential facts that lead directly to confident decisions and their speedy implementation. His fullest approach to success, to happiness, and to a genuine "Brotherhood of Man," as he admitted in this essay, was through his willing engagement as a volunteer in the leadership of four great humanitarian organizations: the American National Red Cross, The National Foundation for Infantile Paralysis, Tuskegee Institute, and the National Conference of Christians and Jews. He was especially proud of these four. It is surprising that he failed to find a fitting place for Dartmouth College in this formulation of his all-embracing altruism, but then the Dartmouth connection was largely social and fraternal. In his final curriculum vitae, these four organizations remained paramount, but most especially did his inner spirit reach its most enduring expression in the creation and advancement of the NFIP, known today as the March of Dimes.

CHAPTER 3

O'Connor and Roosevelt at Warm Springs

Basil O'Connor met Franklin D. Roosevelt (FDR) in 1920 during the campaign that first placed Roosevelt squarely in the national spotlight, when he ran as the Democratic vice presidential candidate under Ohio Governor James Cox. The Democratic ticket lost and FDR's life changed irreparably the following year with the onset of polio that left him a paraplegic for life. Accounts have differed on the date of their first meeting; some have erroneously placed it in 1924 when O'Connor witnessed FDR's struggle to recover from a fall in the lobby of the Equitable Building at 120 Broadway in Manhattan where O'Connor had his law office and Roosevelt worked for Fidelity and Deposit.[1] Jean Edward Smith claims that the two began working together in 1923; they formally announced the law partnership of Roosevelt and O'Connor in 1925.[2] A National Foundation for Infantile Paralysis (NFIP) backgrounder about O'Connor for the occasion of his 60th birthday in 1952 dates their acquaintance from 1920, *before* FDR was stricken with polio.[3]

The law firm Roosevelt & O'Connor commenced auspiciously on New Year's day in 1925, devoted to the "general practice of law," with offices remaining at 120 Broadway, later to become the first headquarters of the NFIP. Junior attorneys included Benjamin F. Crowley, Egbert H. Womack, and Joseph E. Worthington, Jr., all associates of O'Connor. Roosevelt immediately sent his greetings to O'Connor with the warm message that "I want you to know how very happy I am having you as a partner and friend." Widely noted in the press, the *New York Herald Tribune* claimed that Roosevelt had not practiced law in 15 years and that his health was "almost entirely regained" after returning from "the South," a typical assessment of his disability that FDR and advisors like Louis Howe encouraged. Congratulations poured in from all sources. Robert O'Brien of the *Boston Herald* congratulated O'Connor for

---

[1] Friedel, Frank. *Franklin D. Roosevelt: The Ordeal*, (Boston: Little, Brown & Co., 1954) p. 119.

[2] Smith, Jean Edward. *FDR*. New York: Random House, 2007. p. 695, note 60. See also DART, Basil O'Connor Papers, Roosevelt and O'Connor law firm, bound volume; 1925–1933.

[3] MDA, Basil O'Connor Papers, Series 2: Dinner Programs. January 7, 1952.

Friends and Partners
ISBN 978-0-12-803597-9
http://dx.doi.org/10.1016/B978-0-12-803597-9.00003-5

"forming an alliance with the great leader of the Democratic Party," even suggesting that a presidential ticket of Roosevelt and O'Connor might be "invincible." James Forrestal, later to become Secretary of the Navy under FDR, wrote, "From what I know of Mr. Roosevelt you have made an alliance with one of the outstanding personalities of this country."[4] Roosevelt's participation in the law practice created from their compatible ambitions seems more form than substance; O'Connor himself admitted to giving Roosevelt's name priority only for their mutual advantage. During the early years of the partnership FDR was preoccupied with the arduous recovery from polio and the creation of the rehabilitation center at Warm Springs. After Roosevelt assumed the presidency in 1933 the firm was dissolved, to be succeeded by O'Connor and Snyder, and then O'Connor and Farber.[5] At that point, FDR would write to his partner, "Now that I am safely ensconced in the White House I want you to know that the one 'fly in the ointment' is the thought that our partnership must temporarily come to a suspension." Yet their law practice was only one facet of a complex association.[6]

Much has been written about another legendary encounter: that of FDR with the little town of Warm Springs, Georgia, once known as Bullochsville. Here he found not only additional hope for personal recovery but also a way to transform his quest for health into a full-fledged rehabilitative program that eventually led to the eradication of polio. A recent study has depicted Roosevelt's polio colony there as a precursor to the Independent Living Movement in the United States through FDR's visionary precedent of creating an egalitarian community of polio convalescents.[7] O'Connor and Roosevelt's partnership in general was enriched immeasurably by their collaboration on bringing the resort from utter obscurity to the vanguard of the attack on polio, and their friendship matured further as Roosevelt fashioned his reentry into national politics. In the trajectory of Roosevelt's career, O'Connor's counsel was an ever-dependable source of pragmatic guidance; the latter's own transformation into a formidable executive leader developed through

---

[4] DART, Basil O'Connor Papers, Roosevelt and O'Connor law firm, bound volume; 1925–1933.

[5] *New York Times*, March 15, 1976. According to the *Times*, O'Connor met FDR "in 1920 when [FDR] was running for Vice President on the Democratic ticket. And he was, of course, profoundly concerned when Mr. Roosevelt contracted the disease in the summer of 1921." *Ibid.*, March 10, 1972.

[6] DART, Basil O'Connor Papers, Roosevelt and O'Connor law firm, bound volume. Letter from Franklin Roosevelt to Basil O'Connor, March 8, 1933.

[7] Holland, Daniel. "Franklin D. Roosevelt's Shangri-La: foreshadowing the Independent Living Movement in Warm Springs, Georgia, 1926–1945." *Disability and Society*. August 2006. 21(5):513–535. The first popular treatments of this story are *A Good Fight: The Story of FDR's Conquest of Polio* (1960) by Jean Gould and *The Squire of Warm Springs: FDR in Georgia, 1924–1945* (1978) by Theo Lippman, Jr.

his multidimensional association with Roosevelt. Together, in October 1924, the two men visited Warm Springs for the first time, and within 2 years FDR had signed a deed for the purchase of the run-down resort, utterly delighted by the liberating and therapeutic effects of the warm pool waters on his paralyzed legs. FDR wrote to his mother Sara Roosevelt in the autumn of 1924 with a report about Warm Springs. He told her, "I am going to have a long talk with Mr. George Foster Peabody who is really the controlling interest in the property. I feel that a great 'cure' for infantile paralysis and kindred diseases could well be established here."[8] FDR acquired the property on April 29, 1926. So impressed was he by the rehabilitative effects of the naturally warm waters from the spring that had attracted vacationers since the mid-19th century, Roosevelt took the risk of transforming the resort into an aftercare treatment center for polio sufferers. O'Connor expressed his outspoken skepticism about the plan despite Roosevelt's expectant hopes. In a 1946 speech on the history of Warm Springs, O'Connor admitted that his view was totally contrary to his law partner's vision:

> I don't believe I had ever heard of Warm Springs, Georgia, before Roosevelt mentioned it. There were times during the following year, if I may be frank, when I wished I never had heard of it. It was a run-down establishment for which I had no enthusiasm whatsoever. Buildings were falling down, roofs leaked and there was an air of hopelessness about the place that dulled what little optimism I could generate on behalf of its future. … My partner saw it in different light, because one of his finest assets was vision. He could dream practically, and he didn't have to see all four parts of a square closed. If he could see a certain distance ahead, he never grew careless about the rest of the journey, but he didn't worry, either.[9]

At Warm Springs Roosevelt shared with others both his polio disability and his optimistic belief in recovery, creating a social environment that tended to equalize all as it exploded the stigma of disability. However, as Daniel Wilson has pointed out, the hard discipline of rehabilitative treatment was painful and often frustrating, experienced as much by setbacks, disappointment, and exhaustion despite the utopian glow that FDR imparted to the place in his most encouraging pronouncements.[10] Although Louis Howe had at first attempted to shield the true extent of FDR's disability from public view with several cagey stratagems, it soon was made public in several news stories. And while the Warm Springs episode may be

[8] Roosevelt, Elliott, editor. *F.D.R. His Personal Letters.* Volume II, 1905–1928. New York: Duell, Sloan & Pearce, 1950. p. 568.

[9] MDA, Georgia Warm Springs Foundation Records, 1920–1976; Series 1: Correspondence, General Affairs, Basil O'Connor publicity speech; May 1946.

[10] Wilson, Daniel. "The Legacy of Warm Springs." Unpublished talk, Roosevelt Warm Springs Institute for Rehabilitation, 2007.

understood as FDR's personal breakthrough from the effects of isolation and psychological depression—helping himself by helping others—his situation was a shared experience from its onset. In fact, FDR began responding to correspondence from "fellow sufferers" of poliomyelitis within a mere month of his attack after the first reports of his illness in the press. An octogenarian in Minneapolis, Elizabeth Carleton, wrote to him on September 17, 1921 expressing her get-well wishes, revealing that she, at 87½ years, walked with a cane after a bout with polio. Within a week FDR replied:

> If I could feel assured that time could treat me so lightly as to leave me at eighty-seven and a half years with all my vigor, powers and only a cane required, I would consider that my future was very bright indeed. There are not many people who can equal that record, even though they have been fortunate enough not to have been fellow-sufferers, with you and me, of infantile paralysis.[11]

O'Connor argued against full involvement in the resort, insisting that "a hard choice would have to be made right away between vacation spot or medical facility … or both would surely fail."[12] The two purposes clashed from the start, and after complaints from the nonpolio crowd of vacationers, a separate dining room was provided for paralytics. FDR went further by building a small treatment pool "thirty yards from the public one for the use of his 'gang.'" Some residents erected small cottages for lengthy stays. Thus FDR came to be known during these early years at Warm Springs as "Dr. Roosevelt," for it was he who advised the first patients how to exercise in the water. There were no polio specialists at Warm Springs at this time, and FDR undertook the task of teaching what he had learned from Dr. Robert Lovett, who had originally diagnosed his case as polio.[13] The wife of the concert pianist Lee Pattison, who later became a regional director of the WPA Federal Music Project in New York, was one of the first patients. The general tenor of the place was so extremely informal at first, that she had a cottage erected on the property "in order to be able to consider Warm Springs a home."[14] The Pattisons' daughter Tish died there in July 1928, the first to be buried in what became a cemetery at the site.[15]

Neither was Eleanor Roosevelt nor Louis Howe, FDR's long-time political advisor, especially encouraging at the outset of this new enterprise. FDR's son James later wrote that "Howe and O'Connor provided the most help for GWSF, though BOC was reluctant at first."[16] In the continual

---

[11] Roosevelt, Elliott, editor. op cit., p. 529.

[12] Ward, op. cit., p. 715.

[13] Roosevelt, Elliott, editor. op. cit., p. 578.

[14] Ibid., p. 634.

[15] MDA, *The History of the National Foundation for Infantile Paralysis*. Volume II, Book 1. p. 37, n. 191.

[16] Roosevelt, James and Sidney Shalett. *Affectionately, FDR: A Son's Story*. New York: Avon Books, 1959. p. 157.

resurgence of optimism that he would ultimately walk again on his own, FDR spread the gospel of personal rehabilitation to the first of the disabled sojourners who arrived at the new polio center and cast himself not only as "Doc Roosevelt" but as "Vice President in Charge of Picnics." O'Connor, on the other hand, became the treasurer of a "nonexistent treasury," assuming responsibility for an ailing physical plant, resources stretched too thinly, and unpaid bills. The constitutional differences of the two men are portrayed vividly in *The History of the National Foundation for Infantile Paralysis*:

> The tidy O'Connor, meticulously groomed and manicured, wearing the inevitable boutonniere, is hardly the one who would have shared Roosevelt's enthusiasm for the informal fish fries, the corn pone, and the "singing around the dying fire" on Pine Mountain … [Warm Springs] was, to O'Connor, frankly "a miserable mess." It is doubtful he quite outgrew this original feeling. … Besides a natural antipathy to the primitive, O'Connor probably also disliked Warm Springs because to his practical, clear-headed way of thinking, it was a bad bargain, an ill-considered venture in which his friend and new law partner proposed to sink about two-thirds of his private resources. … It was O'Connor's unhappy task to step in after the business deal had been consummated and try to restrain his friend's "leaping enthusiasm."[17]

An Arduous Struggle, c.1926. Franklin Roosevelt, Fred Botts, and an unidentified person using parallel bars to provide support during a therapy session at Warm Springs. This rare photograph of a smiling FDR in therapy belies the constant pain and exhausting effort required to improve muscular strength. Fred Botts became a stalwart friend of FDR and Basil O'Connor; he went on to become the Director of Admissions at Warm Springs.

[17] MDA, *The History of the National Foundation for Infantile Paralysis*. Volume II, Book 1, pp. 75–76.

If their law partnership seemed a mere formality, by comparison the colony at Warm Springs became an endeavor in which the lives of the two were mutually and permanently bound up. Before the "Little White House" was constructed, FDR built his first house at Warm Springs in 1926; it was later occupied by O'Connor.[18] Roosevelt also purchased a nearby farm, and O'Connor would peevishly curse the farm while carefully studying its management and recommending capital improvements, characterizing FDR's fraternizing with the local farmers as part of the Roosevelt "hokum."[19] In truth, the fraternizing provided a critical perspective on the plight of farmers sympathetically addressed by FDR in the very first days of the New Deal. Arthur Carpenter, a polio patient elevated to business manager of Georgia Warm Springs Foundation (GWSF) when FDR became New York Governor, remarked that O'Connor's active involvement in Warm Springs "conformed to the law of uniform acceleration." Carpenter wryly explained that up to around the time of Roosevelt's presidency, O'Connor operated in "low gear"; during the mid-1930s, he ramped up to "second gear"; and when the NFIP was founded in 1938 he went into "high gear" and stayed there without downshifting ever again.[20] In the end Basil O'Connor's interest in the endeavor endured, outlasting Roosevelt's, and equally momentous.

Despite Roosevelt's do-it-yourself approach to polio rehabilitation at Warm Springs sketching his own muscle charts and directing exercises in the pool, he had heeded the advice of several expert physicians. However, to bring a stamp of medical legitimacy to the enterprise at Warm Springs Roosevelt petitioned the American Orthopedic Association at their national convention in Atlanta to form an investigating committee to explore the worthiness of hydrotherapy rehabilitation at Warm Springs. The physician in charge of the committee was Dr. LeRoy W. Hubbard, an orthopedic surgeon of the New York State Department of Health, with extensive experience in treating polio patients. Hubbard designed and conducted an experiment involving the observation of 23 patients in hydrotherapy treatment from June to December 1926. His findings were generally positive, and with his endorsement the committee recommended "the establishment of a permanent hydrotherapeutic center at Warm Springs."[21] In January 1927, FDR formed the GWSF as a "nonstock, nonprofit institution" on the basis of Dr. Hubbard's

[18] RWSIR. Curry, James E. and Rexford G. Tugwell. "Franklin D. Roosevelt's Visits to Warm Springs, 1924–1945," typescript. April 8, 1958.
[19] Tugwell, Rexford G. *The Brains Trust* (New York: Viking, 1968) p. 329.
[20] MDA, *The History of the National Foundation for Infantile Paralysis*. Volume II, Book 2. p. 80.
[21] Roosevelt, Elliott, editor. op. cit., p. 610.

report and recommendation. The incorporators were "Doctor" Roosevelt, George Foster Peabody, Basil O'Connor, Herbert N. Straus, and Louis Howe. Dr. Hubbard became the chief physiotherapist and director of nurses. FDR wrote to his Aunt Bye "I am sending you some of our folders about Warm Springs. The work of starting a combined resort and therapeutic center has been most fascinating for it is something which, so far as I know, has never been done in this country before." FDR continued, waxing romantic about the place: "Oh, I do wish that you could be wafted down there and placed gently in a chair and slid gracefully down a ramp into the water. You would love the informality and truly languid southern atmosphere of the place!"[22]

Under Roosevelt's direction the facility developed over the years from a primitive hydrotherapy center to a well-equipped medical care facility with hospital, pools, dormitories, infirmary, chapel, and school. A brace shop created customized orthopedic devices, and trained professionals managed a convalescent care program in physical and occupational therapy for children and adults disabled by polio. Basil O'Connor first served as Treasurer and Chairman of the GWSF Executive Committee; during the war years in the 1940s he assumed the roles of both Vice President and Treasurer. On occasion, more frequently during the war, FDR appointed O'Connor as his proxy for board meetings.[23] O'Connor, while remaining Treasurer, succeeded FDR as President of GWSF after FDR's death in 1945, and John S. Burke of B. Altman & Co. assumed the chairmanship of the Executive Committee. At the establishment of the NFIP in 1938, the two foundations remained separate entities, and O'Connor seamlessly juggled the responsibilities of leading each. The NFIP awarded grants to GWSF and funded building projects such as the construction of the Roosevelt Hall rehabilitation center. Local March of Dimes chapters often financed Warm Springs' patient care. In 1941, GWSF appointed Dr. Robert Bennett as Director of Physical Medicine and established postgraduate education in physical and occupational therapy. Later, with the advent of the Salk vaccine and mass immunization against polio, Dr. Bennett, as Executive Director, revamped its program to include physical and internal medicine and orthopedic surgery in the treatment of neuromuscular and muscular–skeletal disabilities resulting from birth defects, trauma, or disease. In 1974, GWSF sold its hospital and associated properties to the State of Georgia, which renamed it the Roosevelt Warm Springs Institute for Rehabilitation in 1980. At the present writing it remains an accredited medical and vocational rehabilitation center.

---

[22] Roosevelt, Elliott, editor. op. cit., pp. 618–19, 623, 624.
[23] MDA, Basil O'Connor Papers, 1931–1984; Series 4: FDR, letters of proxy.

King and Queen of the Day, c.1928. FDR (right of center) waves a white handkerchief to mark the opening of the Ford pool at Warm Springs for a swimming party known as "King and Queen of the Day." The pool was used for hydrotherapy as well as for water play and sporting events by the patients at Warm Springs, including Roosevelt himself. Swimming and water therapy were central to FDR's concept of polio rehabilitation. He once confided to Basil O'Connor, "I am a crank on keeping the pool open."

Prior to his reentry into political life as Governor of New York in 1928, Roosevelt placed the stamp of his personality in the early operations of the center, mingling freely with the patients and giving medical advice as a self-appointed lay "doctor." In fact, he cultivated the persona of "Doc Roosevelt," immersing himself in the lives of fellow "polios," as these patients were called at the time, just as he was immersed with them literally in the buoyant waters of the pools. FDR cultivated key associates from all walks of life to assist him in the endeavor, and many became essential staff members. Dr. LeRoy W. Hubbard was the medical director until 1931 as Dr. Michael Hoke replaced him when a complete medical and surgical service was initiated. Fred Botts arrived one day for treatment himself and found a lasting role as Registrar, forever befriended by FDR. Arthur Carpenter became the business manager, and Alice Lou Plastridge was the chief of physical therapy. In the physical layout of the property FDR wanted a patient's experience to feel more like that of a college campus or a country club than a hospital. GWSF was in fact modeled on the campus of the University of Virginia. In keeping with this, the first board of trustees stipulated, "It is *not* the policy of the trustees to develop Warm Springs into *either a hospital* or a *sanitarium*"[24] (original emphasis). It was called "the colony" from its appearance as a residence, not an institution. It is important to realize that GWSF grew incrementally over time; the first instance of major expansion was the construction of a glass-enclosed swimming pool for patients in 1928 thanks to a gift from Edsel Ford, son of the industrialist Henry Ford. The Norman Wilson Infirmary was erected in 1930 at a cost of $40,000 while Georgia Hall, the central administration building, was completed in 1933 after a special fund-raising campaign. Dormitories and other buildings followed, a chapel was built in 1937, and a complete orthopedic hospital in 1939.[25]

As the executive head of GWSF, Roosevelt concerned himself with all aspects of the operation until his decision to run for governor of New York in 1928. Roosevelt had already delegated much of the daily management by then, but at that point O'Connor became the de facto director, doing everything from discharging poorly performing employees to negotiating utility power rates. Attentive to publicity as well as development, FDR issued an illustrated pamphlet, "A Pioneering Opportunity," that set forth a rationale and credo in the form of an appeal: "I think most cripples, children or adult, are worth taking an interest in. Economically, this work is sound; humanly, it is right. Incidentally, it is reaching out into a field which no other agency is now adequately reaching. We need pioneers."[26] FDR's "Shangri-La" at Warm Springs indeed established a pioneering social environment where disability was not stigmatized as it was in society at large and "where the medical model of disability was repudiated."[27] The pain and suffering of Roosevelt's own recovery from polio was sublimated through his polio humanitarianism, though the cheerful naiveté of the Warm Springs catchphrase "happy though handicapped" belies the excruciating struggles experienced by so many of painfully retraining flaccid muscles into a semblance of functionality.[28]

---

[24] MDA, Georgia Warm Springs Foundation Records. Series 1: Correspondence. Little White House dedication ceremony, 1947.

[25] MDA, History NFIP Records; Series 2: Typescripts. Roy A. Chambliss, "A Social History of the NFIP, Inc., 1938–1948." New York: Fordham University School of Social Service, 1950.

[26] MDA, Georgia Warm Springs Foundation Records; Series 3: Typescripts. Lunoe, Adam M. "Notes on the Georgia Warm Springs Foundation." 1947, p. 18.

[27] Holland, Daniel. op. cit.

[28] RWSIR, "Hope and Courage…That is Warm Springs," 1944. See also Wilson, Daniel. "And They Shall Walk: Ideal versus Reality in Polio Rehabilitation in the United States," *Asclepio: Revista de Historia de la Medicina y de la Ciencia* 61(2009): pp. 175–92, and "A Crippling Fear: Experiencing Polio in the Era of FDR," *Bulletin of the History of Medicine* 72 (1998), pp. 464–95.

By 1931, as the nation slipped deep into the destitute hopelessness of the Great Depression, Roosevelt began to attack the problems of economic malfunction and unemployment as Governor of New York. Submerged as he was in the duties of the governorship and positioning himself for a run for the presidency, he managed to visit GWSF several times while governor. As for his own physical condition as the campaign for the presidency took shape, Dr. Hubbard attested that "[a]side from the weakness of the muscles of his legs, due to the attack of poliomyelitis which he had in 1921, he has been in perfect physical condition."[29] This pronouncement was far removed from the full truth, and Roosevelt would continue to present his disability outside the confines of Warm Springs under the guise of nearly complete recovery to validate the doctor's words as routine and his paralysis as inconsequential. As such, Roosevelt was commonly perceived as "the cured cripple." In July 1931, a patient publication called *The Polio Chronicle* was launched at Warm Springs which carried one of Roosevelt's most incisive statements about polio. *The Polio Chronicle* enthusiastically congratulated FDR in its "Founder's Number" of December 1932 for his successful bid for the presidency; in turn, Roosevelt wrote:

> Can we, who have felt the spirit of Warm Springs, do less than our best in a crusade launched against the vast human loss inflicted by infantile paralysis? We all have glimpses, however vague, of the whole picture; we understand at least a little of the human values involved in the stupendous problem presented by hundreds of thousands of polio victims. We have a gospel to preach. We need to make America "polio conscious" to the end that the inexcusable case of positive neglect will be entirely eliminated.[30]

In a patient publication this was certainly preaching to the choir, but it was also an ex post facto campaign promise that claimed enormous stakes in the direction toward which the "polio crusade" was now cast. *The Polio Chronicle* had characterized itself as the "Official Organ of Polio Crusaders, a National Movement against Infantile Paralysis," and with this aspiration the ideal of an expansive social movement against polio moved closer to realization. "Polio consciousness" had received an impressive boost now that "Polio Crusader Number one has become Citizen Number one."[31] But before FDR could fix polio, he first had to fix the nation. It was not a propitious moment to rally the millions to the polio cause, but 1 year later, after the essentials of the New Deal had begun to rejuvenate an ailing economy, the nation was ready.

---

[29] FDRL, Roosevelt Family Papers Donated by Children. Basil O'Connor, Box 24.

[30] *The Polio Chronicle*, December 1932.

[31] MDA, Georgia Warm Springs Foundation Records. Series 3: Typescripts. Lunoe, Adam. "Notes on the Georgia Warm Springs Foundation," p. 24.

Board of Trustees, Georgia Warm Springs Foundation, January 1931. [back row, left to right] Paul G. Richter, George Foster Peabody, Basil O'Connor, Arthur Carpenter, Frank C. Root; [seated, left to right] Dr. LeRoy W. Hubbard, FDR, Leighton McCarthy. Dr. Hubbard was an orthopedic surgeon who served for many years at GWSF, and Arthur Carpenter was its general manager. Leighton McCarthy, a Canadian business leader whose son had polio, was a stalwart supporter for many years. The philanthropist George Foster Peabody, who owned the property at Warm Springs before Roosevelt, had first invited FDR there to test its hydrotherapeutic waters.

The inauguration of Franklin Delano Roosevelt to the presidency on March 4, 1933 was a momentous occasion, and Americans from coast to coast had a keen realization of its gravity as they tuned their radios to hear FDR intimate his purpose, and his plans, for a nation shaken to its very core by unemployment, idle factories, barren farms, civil unrest, starvation, and bank closings. His assertion in his inaugural address intertwining the themes of fear and paralysis hit home immediately and stands today as one of the most memorable utterances of *any* American president: "So, first of all, let me assert my firm belief that the only thing we have to fear is fear itself—nameless, unreasoning, unjustified terror which paralyzes needed efforts to convert retreat into advance." In that time of crisis, the public awareness that FDR personally understood both paralysis and fear, and the reactive fear provoked by bodily paralysis, was widespread and unquestioned, notwithstanding any retrospective judgment that he hid or disguised his polio disability as a form of psychological denial. Roosevelt "stage-managed" his disability.[32] The theatricality of his performances in presenting his disability as *past* misfortune that he had overcome became an essential tactic in humanizing his image, an image which O'Connor would consistently seek to align with the ideology of charity and American patriotism. His inability to stand or walk on his own was variously put on display (at Warm Springs, at Hyde Park, and at the White House), suppressed by buoyant optimism (his generous smile), diverted from attention (his love of fun, sports, and play), carefully regulated for the press (no photos while in his wheelchair), dismissed as irrelevant (his qualifications for public office), and candidly shared with others (news stories about his convalescence and his own letters to the disabled). His articulation of the fact that *fear paralyzes* and his subsequent directives in the first "one hundred days" of the New Deal demonstrated that the nation had turned a corner, if ever so slightly, toward recovery. FDR proved that on the political stage he was compassionate but fearless, and the cultural myth of his being "cured" of polio persisted, to be modified as needed in the campaigns to come. FDR and his advisors were brilliantly successful in refocusing attention to his magnanimity of character and infectious optimism expressed in his actions.

As an invited guest to the inauguration, Basil O'Connor received from the Inauguration Committee a deluxe hardbound inauguration program with his name embossed on the cover; this copy is preserved today in the Basil O'Connor Papers of Dartmouth College. Doc's copy of the book contains many scribbled autographs of those in his private party: Rexford Tugwell, law partner William F. Snyder ("a hard-drinking Dutchman"), "Hoopy the Irishman," and several others of all of whom were brimming with jubilation.[33] The written précis of FDR in the program is revelatory in its summary of his career thus far. Particularly interesting is the equivocal portrayal of Roosevelt's bodily strength and his infirmity; the program stated: "Franklin D. Roosevelt comes to the Presidency in his fifty-second year, over six feet tall, with the torso of a heavy-weight wrestler, all his hair, and a smile that has become famous. His life has been a battle ..." After touching

---

[32] Moeschen, Sheila C. "Benevolent Actors and Charitable 'Objects': Physical Disability and the Theatricality of Charity in Nineteenth- and Twentieth-Century America." Northwestern University, Ph.D. dissertation. 2005.

[33] DART, Miscellaneous correspondence and memorabilia. Inauguration program, Franklin D. Roosevelt and John N. Garner. March 4, 1933.

on early political battles, his struggle with polio was acknowledged: "It was in 1921 that he faced his hardest and greatest battle, for he was attacked by what laymen call 'infantile paralysis,' a germ horror which nearly killed him and left his legs partially atrophied. Although he suffered an attack which would have rendered the average person a helpless cripple, he practiced law successfully and weathered two strenuous and victorious campaigns for the Governorship of New York …"[34] Here, in this public document marking the inauguration of the President, the two sides of Franklin Roosevelt are expressed in a stunning contradiction: he embodies the attributes of both a "heavy-weight wrestler" and a "helpless cripple." That his polio disability was life-changing but "*would have rendered the average person a helpless cripple*" strongly suggests that he might indeed be considered a "helpless cripple" but that he was most assuredly not an "average person." The two terms negate each other; the lasting result of his encounter with polio is glossed over; and FDR is revealed as simply a man.

Yet, Roosevelt was a political man, and his image as a leader wise enough to lead the nation out of the abyss had to be tended carefully. His image in political cartoons consistently characterized his physical strength and fortitude; these were certainly welcome but completely beyond his control. Basil O'Connor collected these cartoons assiduously. But O'Connor did something more: he advised that FDR forgo any further mention of travel to Warm Springs for the purpose of convalescence, relaxation, or vacation. From here on, O'Connor told him, any announcement about Warm Springs must focus strictly on FDR's administrative business there only—yet another point of stage-managing his disability.[35] Before long, even though his personal valet needed to assist him with bodily functions and to help get his clothes on every morning, the "helpless cripple" imagery began to recede even further from the public eye as FDR became a fixture in the news media as the eminent leader of the nation's economic recovery. Random House publisher Bennett Cerf described a meeting with FDR in 1937 that shows just how far the popular image of FDR had jumped from disability to capability:

> Sam Rosenman drove me up to Hyde Park on July 3, 1937. It was an exciting trip: I was going to meet one of my heroes, President Franklin D. Roosevelt! … Out came the President, wheeling himself. He manipulated that wheelchair like a racing driver. The house at Hyde Park had special ramps and he wheeled up and down like a bat out of hell. As he came out, my heart jumped at the sight of the President of the United States waving his hand; when he said "Hello, Bennett," I was ready to let

---

[34] DART, ibid.
[35] MDA, *The History of the National Foundation for Infantile Paralysis.* Volume II, Book 1, p. 31, note 159.

him roll over me. We went in to lunch after a little while, and he had to be lifted into his chair at the table, which was a shock; but the minute he was seated, he was President Roosevelt. It was only his legs that were bad, and as he sat at the table I forgot immediately that he was a cripple. He was completely in charge and utterly and totally charming. When he wanted to be, he was irresistible.[36]

The onerous duties of the American presidency took Roosevelt away from the sense of exhilaration and transformative purpose that he experienced at Warm Springs. He had spent two-thirds of his own fortune to purchase the property, yet the colony led a hand-to-mouth existence and fund-raising from the beginning was catch-as-catch-can. Roosevelt and O'Connor each donated or loaned from their own finances periodically to meet critical needs. Early fund-raising was by subscription, and O'Connor usually mailed personal appeal letters himself until Keith Morgan came on board to manage fund-raising. For example, the Baltimore banker Van Lear Black promised $18,000 in 1929, and he made four quarterly payments of $4,500.[37] But the next year, as the financial situation at Warm Springs mirrored the American economy after the crash, and with revenue at only a trickle, O'Connor recommended that the foundation keep top donors like Edsel Ford apprised of its situation in anticipation of seeking major gifts.[38] The economic downturn hit Warm Springs hard. Victor M. Cutter, President of the United Fruit Company, answered one of O'Connor's appeals, "I have your letter of April 11 and realize the difficulty you must have had with the Warm Springs Foundation. Personally, I am just as flat as the Foundation possibly could be. Nevertheless, this is one thing that I sincerely believe in. Therefore I am enclosing [a] check for my last hundred dollars."[39] In these circumstances such generosity was fortuitous; more frequently the responses ran like this one: "Thanks for the compliment of asking me for $1,000. I haven't $1,000, Doc, and if I did I think there are many cases [at home] that could use a little money very nicely."[40]

As the Depression deepened, the foundation preferred to spend its meager funds on patient aid rather than new construction projects. In 1931, the patient aid fund was still rather small ($17 per week) leaving the amount to be paid by a family at $25 per week. The purpose of the "polio crusade"

---

[36] Cerf, Bennett. *At Random: The Reminiscences of Bennett Cerf.* New York: Random House, 2002. p. 138.

[37] NYSA, Basil O'Connor Papers. Van Lear Black to O'Connor; December 27, 1928.

[38] FDRL, Roosevelt Family Papers Donated by Children. Folder 7, Basil O'Connor. January 20, 1930.

[39] NYSA, Basil O'Connor Papers, Victor M. Cutter to O'Connor. April 13, 1932.

[40] Cutlip, Scott M. *Fund Raising in the United States.* New Brunswick: Transaction Publishers, 1990. p. 356.

launched in 1932 as a "national movement" was therefore twofold: to educate the public about polio and to raise money for those unable to afford GWSF admission rates to finance their stay at Warm Springs. The drive to provide for the foundation's patient aid fund was thus one stimulus for a broad-based appeal that prefigured the aims of the later March of Dimes campaigns to finance patient aid on a national scale. The fund had been administered by a committee of trustees for patients who were unable to pay the cost rates. The National Patients' Committee organized at Warm Springs in 1931 intended to make America as "polio conscious" as the Christmas Seals program had made the country "tuberculosis conscious."[41] Membership in the crusade was $1 by subscription, managed by the treasurer of the committee. Roosevelt's inauguration sparked a new hopefulness about improving the financial situation of Warm Springs. Letter-writing campaigns to philanthropic friends continued as before, but Roosevelt's unquestioned popularity was the vital factor that sparked a new approach in the early months of 1933, even though the situation failed to turn around quickly. Keith Morgan assumed leadership of another letter-writing appeal that came to be known as "Help the President," premised on the idea that contributions would relieve FDR of the responsibility of the financial management of Warm Springs. A benefit concert at Carnegie Hall in New York City featuring Lee Pattison was scheduled for December 8, 1933, and between that time and FDR's birthday on January 30, a new plan was hatched that had not only immediate results but long-term implications for the future of the foundation and the ultimate defeat of polio.

In its January 1934 issue *The Polio Chronicle* gleefully announced: "Birthday Balls Everywhere."[42] The "birthday balls" associated with FDR were annual fund-raising celebrations held in communities across the United States to benefit GWSF. This first occurred on January 30, 1934. Keith Morgan, the fund-raiser for GWSF, developed the idea for the event as a short, concentrated campaign, and FDR gave his consent to use his birthday. The first birthday ball was not a single event but took place in 4,376 communities in 6,000 separate celebrations. The *Detroit News* reported that the celebrations were marked by "merriment, gaiety, and good cheer as young and old, rich and poor, danced the hours away." Marion Clausen, a coauthor of the *History of the NFIP* saw an even more poignant meaning in the celebrations: "It seemed as if this occasion signaled the emergence from the hard times, the gloom, the necessarily restricted gaiety of the long depression years.

---

[41] *The Polio Chronicle*, August 1931.
[42] *The Polio Chronicle*, December 1932 and January 1934.

The pent-up desire for fun, for frills and fancy dress, for ornate and elaborate parties, for music and pageantry was given a chance to express itself."[43] The Committee for the Celebration of the President's Birthday was organized to manage the event each year; Keith Morgan became its National Chairman. Birthday ball galas became a tradition on January 30 through the 1930s at venues ranging from a homely dinner-dance in a potato warehouse to high society events at New York's Waldorf Astoria and Chicago's International Amphitheater. The slogan for the event—*Dance, That Others May Walk*—was crafted into a tune by George M. Cohan and sung by Lily Pons. Morgan engaged the services of Colonel Henry L. Doherty, having heard of his personal struggle with arthritis, and set out to convince him that he should join the polio cause. Doherty brought on Carl Byoir, whom Morgan knew as a boy when he worked for William Randolph Hearst. Morgan introduced Doherty and Byoir to FDR at the White House in the summer of 1933. They then attended the next Founder's Day dinner on Thanksgiving at Warm Springs, afterward seeing the president frequently and developing plans for fund-raising. Morgan insisted on a major national effort, Byoir spoke up in favor of a one-day affair, and the President agreed that his birthday would be wholly appropriate. Morgan then made a whirlwind trip that culminated in the first birthday ball: "After a feverish organization campaign I made a lightning trip through the key cities in the Winnie Mae [airplane] with Wiley Post to select the chairmen and spark plug the drive."[44]

The success of the first birthday ball was surely a measure of the President's popularity, at its peak after 10 months in office. The proceeds totaled over $1 million, and Basil O'Connor professed his utter amazement at the amount raised, a windfall far beyond his or anyone's expectations, especially in the economic crisis. The *New York Times* commented that FDR was "proving to be a veritable Pied Piper to the nation's crippled children."[45] However, the subsequent political charges that FDR was lining his own pockets derived from questions about the enlistment of Democratic Party officials, newspaper publishers, mayors, and US postmasters to help organize thousands of local events. One writer has claimed that Byoir approached the Democratic Party chairmen in each community or "the Roosevelt-appointed postmaster." In reality, many postmasters who had taken the

[43] MDA, *The History of the National Foundation for Infantile Paralysis.* Volume II, Book 1, p. 63.
[44] MDA, History NFIP Records, Series 2: Typescripts. Keith Morgan to Basil O'Connor, "How the NFIP Came into Being." 1939.
[45] *New York Times.* "Cripples Pin Hope to Roosevelt Aid," quoted in The History of the National Foundation for Infantile Paralysis. Volume II, Book 1, p. 90.

initiative to raise money in local communities were Republicans, and Morgan's reliance on their logistical capabilities varied over the next four years. But in no fashion did Roosevelt, O'Connor, or the Birthday Ball Committee deputize all the nation's postmasters to organize the collection of monies for Warm Springs every January. Doherty himself brought a larger agenda to the President, proposing to link the birthday balls each year to different causes yet carping about Roosevelt's attention to "the redistribution of wealth" in the first year of the New Deal. This brought an immediate rift between him and O'Connor and Morgan, who would never deviate from their commitment to Warm Springs and its focus on polio rehabilitation. But when O'Connor suggested that Roosevelt reimburse his own loans right off the top and keep the remainder of the 1934 proceeds for the Foundation, Roosevelt was furious. Arthur Carpenter reported, "He pounded the desk, something unheard of for him, he shouted his refusal and pointed out that he neither would nor could take any of this 'public money!'"[46]

After the birthday ball monies had been received, tallied, and audited, the Birthday Ball Committee arranged with the White House to schedule a formal presentation of the check for $1,003,030.08 to President Roosevelt. Four hundred invited guests, all of whom had figured prominently in organizing the birthday ball parties across the United States, assembled in the East Room for the presentation ceremony on May 9, 1934. Colonel Doherty was ill and unavailable, but Carl Byoir introduced a retired admiral of the US Navy, Cary T. Grayson, to present FDR with the check. Grayson said, "This gift is the outpouring of the heart of America. It is tendered by friends throughout the nation for a twofold purpose—to help in the great fight against infantile paralysis and as an evidence of the people's appreciation of your great humanitarianism."[47] FDR preferred to read a speech of acceptance rather than rely on extemporaneous remarks since his words would be heard throughout the nation by radio, but he couldn't help adding with a laugh, after he turned the presentation check over to Arthur Carpenter, who then gave it to Treasurer Basil O'Connor, "I am going to appoint you all a committee of the whole to watch 'Doc' O'Connor." The publicity generated by this event was enormous, and it worried O'Connor and even more so Dr. Michael Hoke, who were faced with the dilemma of limited resources versus popular demand. A flood of letters for assistance arrived at Warm Springs in the subsequent weeks and "gate-crashers" soon followed

---

[46] MDA, *The History of the National Foundation for Infantile Paralysis*, Vol. II, Book 1, p. 92, note 207; see also Cutlip, "FDR, Polio, and the March of Dimes," *op. cit.*, pp. 351–96.

[47] Ibid., p. 86.

in the fall "trooping for attention from remote places." A $1 million check, as welcome as it was, could not solve all of Warm Springs' problems, and Hoke was distressed that the overenthusiastic publicity, which he styled "the terrific amount of blatancy" in the newspapers, seemed to exaggerate the capabilities of Warm Springs to the detriment of any standing it might have with the medical profession. As Surgeon-in-Chief at GWSF, Dr. Hoke was rightly proud of what Warm Springs was doing to help "stricken people," but that it seemed on the other hand "the most godforsaken place in its estimation of what to say to get the confidence and esteem of the medical profession."[48]

And yet the gala events continued for the next 4 years up to (and including) the creation of the NFIP and the "March of Dimes" in 1938. They would continue through the war years, merging with the March of Dimes campaigns through 1945, the year of FDR's death. The second (1935) and third (1936) years of the birthday balls suffered a bit from repetition, but each year there were volunteers for the polio cause who brought forward novel ideas. Minneapolis and Chicago both elected a "Queen of the Ball" while in Detroit "matinee dances" in public schools gave rise to a series of public lectures on polio. At the International Amphitheater of the Union Stockyards in Chicago, which the *Detroit News* called an "informal palace of patriotic and charitable whoopee," there were a dizzying array of events, a veritable carnival: boxing matches, side shows, barn dance troupes, cabaret shows, parades, and dice games. Across the country there were "pageants, military drill squads, bands and orchestras, bridge, dancing, vaudeville acts; the President's favorite tunes were played; cakes, some of them weighing over a ton, were auctioned slice by slice to bring sometimes nearly $500."[49] Despite the favorite themes, grassroots organization of these events took on a local character, and the Birthday Ball Committee helped to establish the overarching pattern with the selection of chairmen, especially in larger cities. During this period, there were roughly 5,000 to 7,500 local chairmen; some continued in this role year after year, others only once or twice. The success or failure of the fund-raising depended on these key positions. Most attempted to navigate the political waters by maintaining a strictly nonpartisan agenda with a focus on the humanitarian cause of polio relief, but it was virtually impossible to remove these activities entirely from their association with the President. Still, despite criticism that inevitably arose, local chairmen ran these events with no political scheming or

[48] Ibid., p. 88.
[49] Ibid., pp. 139–40.

"behind-the-scenes partisan manipulation but rather a normal, practical way of handling a problem." This did not prevent elected leaders, such as Georgia Governor Eugene Talmadge, to attack both Roosevelt and the GWSF with wild accusations about exorbitant patient rates and that birthday ball funds were used to retire the mortgage on the property owned by Sara Roosevelt, but this did little to stifle the overall popularity of the cause.[50]

It was inevitable that "politics" came into play repeatedly in succeeding years; as a negative factor, it pushed O'Connor and Roosevelt to seek other means to enlarge the scope of their mission. Critics sniped at FDR, the Birthday Ball Committee, and the Democratic Party itself for the seemingly partisan nature of fund-raising; but such criticism only set the stage to establish another methodology on a nonpartisan basis for polio fund-raising and research. This was one impetus giving rise to the establishment of the NFIP. The social revolutionary nature of birthday balls aligned with the transformation initiated with the New Deal, arousing great interest in the plight of those with polio. In 1934, all the proceeds went to GWSF; in following years they were shared with local communities, except in 1938, when all birthday ball proceeds were earmarked for the NFIP. Yet the money raised each year fell short of what was needed for operating expenses, expanded services, and patient aid at Warm Springs. FDR expressed his concern to O'Connor in a letter of August 9, 1935. Although he was fully confident in his partner's management of the center, he worried about the financial viability of the foundation, especially in carrying so many "charity cases." He confided to O'Connor that "I am applying to the Warm Springs Foundation the same planning that I am using on the budget of the National Government,"[51] implying that fiscal stimulus to the foundation's ailing treasury would be realized in bold, forthright action.

Roosevelt traveled to Warm Springs 38 times after his first visit on October 3, 1924. At Warm Springs he formed the resolve to transform his own polio disability into an institutional means of giving vital support to others in their own recovery and to transform it yet again into a social movement that would banish polio forever. In addition, it was here where he personally witnessed the despair created by economic depression, for hard times arrived in Georgia in the 1920s, years ahead of the general economic collapse of the 1930s. The presidency of the GWSF was the only outside office that FDR retained after his election as US President, though

---

[50] Ibid., p. pp. 139–40.
[51] Joseph Plaud collection of FDR memorabilia. August 9, 1935.

Basil O'Connor assumed the executive leadership of the polio center through FDR's four terms in office. Rexford Tugwell claimed that O'Connor's assumption of the burden of running GWSF was one of the "immense services" that he provided to FDR in his later years.[52] Amid the hardships of the depression and the fervent hopes and changes brought about by the New Deal, O'Connor and Roosevelt understood full well the limitations of Warm Springs and began to collaborate on a greater plan. The seeds of this plan were evident in Roosevelt's mind as early as 1926, the year he founded GWSF. Even then FDR foresaw the possibility of a national organization for he insisted on the name "Georgia Warm Springs Foundation" as opposed to an eponymous designation such as "The Roosevelt Foundation for Infantile Paralysis." Roosevelt admitted, "I want to reserve my own name for an even bigger project which may some day develop."[53]

[52] Tugwell, Rexford G. op. cit., p. 91.
[53] MDA, *The History of the National Foundation for Infantile Paralysis*. Volume I: National Administration and Policies of the NFIP, p. 4.

CHAPTER 4

The Brain Trust

Political insiders during Franklin D. Roosevelt's (FDR) two terms as Governor of New York often referred to Basil O'Connor as "the eyes and ears of FDR in New York."[1] Beyond their collaboration at the Warm Springs colony, Basil O'Connor's knowledge of contracts, utilities, industry, and banking was of inestimable value to FDR. Based in Manhattan, O'Connor became a political intimate of Governor Roosevelt when FDR relocated to Albany in 1928. O'Connor made frequent trips to the state capital for meetings and informal gatherings and gave constant attention to both the law partnership and the financial management of Georgia Warm Springs Foundation (GWSF). With the onset of the Great Depression, FDR was unique among the nation's governors to create an agency that brought temporary relief to the unemployed. In 1932, President Herbert Hoover's inability to change the course of the Depression virtually assured a Democratic victory in the presidential race, and FDR declared his candidacy. This decision brought O'Connor deeper into the governor's circle, though as an unaffiliated advisor he was quite removed from the public eye. The nucleus of men who emerged as FDR's closest advisors on planning and policy in the race for the Democratic nomination came to be known as the "Brains Trust" or "Brain Trust." The group provided critical guidance to Roosevelt in the months before the nomination, through the presidential contest, and into his first term as President. A few, such as Rexford Tugwell, an agricultural economist from Columbia University who became director of FDR's Agricultural Adjustment Administration, and Hugh Johnson, Bernard Baruch's protégé who led the National Recovery Administration, had extremely high-profile roles during the first New Deal.

In his memoir *The Brains Trust* (1968), Rexford Tugwell acknowledged Roosevelt's speech-writer Samuel Rosenman and Basil O'Connor as the "founders" of the illustrious Brain Trust circle, "perfect examples of devoted but undemanding friendship."[2] O'Connor himself initiated

---

[1] DART, Basil O'Connor Papers. General Correspondence, news clippings.
[2] Tugwell, Rexford. *The Brains Trust*. New York: Viking, 1968. p. 28.

Friends and Partners
ISBN 978-0-12-803597-9
http://dx.doi.org/10.1016/B978-0-12-803597-9.00004-7

contact with both Tugwell and Raymond Moley, a leading political economist and law professor at Columbia University. In his book *After Seven Years* (1939), Moley described how he and O'Connor then began to canvass candidates for Roosevelt to interview, rushing to Albany from Manhattan on the late afternoon train, spending an evening of intense discussion at the state capital, and dashing back on the midnight train. O'Connor's workaholic propensity is certainly evident here, for this regime occurred on countless occasions after a full day at the law office. During interviews of potential advisors with Governor Roosevelt, O'Connor was one who intuitively knew which of the meetings were simply informative and which were "deep," best serving Roosevelt's purposes in the upcoming presidential race. Among these men O'Connor also borrowed a certain prestige from that of his brother, John J. O'Connor, a Republican congressman.[3] O'Connor and Rosenman cautioned Tugwell and Moley in the starkest terms about Eleanor Roosevelt's influence on FDR, and the two newcomers were left to figure this personal angle into their service as advisors. O'Connor deemed Mrs. Roosevelt "dangerously idealistic" and often resented her taking the lead about issues in their briefings and dinners together.[4] This would later change as his experience in leadership matured. Eleanor's relationship with O'Connor was yet to ripen and blossom, just as hers had with Louis Howe during FDR's initial crisis with polio, for O'Connor well knew the vital role she had played in FDR's recovery. Preparing the Governor for the coming election was the leading issue at this juncture, superseding all other considerations, even the dire economic issues of the Depression. Howe, Rosenman, and O'Connor knew as almost no one else did that all other issues were subordinate to getting Roosevelt into office, believing that "what was good for the country," in essence, meant "what was good for Roosevelt." These three were among the very few who could contradict and firmly say "*no*" to FDR because, though they were intimate advisors, they operated adjacent to the arena of politics, that is, they were not political stakeholders. Because of this, O'Connor sometimes experienced personal disconnectedness and expressed his opinions cynically, which Tugwell noticed from the first. This deportment as political insider testing the issue philosophically was contrary to the matter-of-fact realism that he endorsed and projected in his capacity as an attorney.[5]

[3] Tugwell, Rexford. op. cit., p. 235.
[4] Lash, Joseph P. *Eleanor and Franklin*. New York: Norton, 1971. p. 346.
[5] Tugwell, Rexford. op. cit., pp. 54, 62, 186–87, 222.

Garden Party for Governor Roosevelt; August 28, 1932. Basil O'Connor holds forth as FDR signs an autograph for Bettyann O'Connor at a garden party at the O'Connor home in Westhampton Beach, Long Island. The news media reported that former Governor Alfred E. Smith declined an invitation to attend the party in a move to distance himself from Roosevelt. [left to right]: Basil O'Connor; his mother Elizabeth Ann O'Connor; John H. McCooey, Brooklyn Democratic leader; Governor FDR; Elvira O'Connor (at FDR's shoulder); Bettyann O'Connor, daughter of Basil and Elvira.

Historical perceptions of the actual membership of the Brain Trust have varied. One writer maintained that the "original Brain Trust" was the foursome Basil O'Connor, Louis Howe, Steven Early, and Marvin McIntyre.[6] However, Early and McIntyre occupied official White House posts as press secretary and presidential secretary, respectively, that is, they were not free-floating advisors like O'Connor (despite the intensive collaboration with the single focus of polio at Warm Springs). According to Tugwell, the core group originally consisted of himself, Raymond Moley, and Adolf Berle (Professor of Corporate Law at Columbia Law School) along with the stalwarts Doc O'Connor and Sam Rosenman. The publisher Robert K. Straus, Hugh Johnson (to become head of the National Recovery Administration), and manufacturer Charles Taussig were "associates." FDR referred to the group as his "privy council" until the term "Brains Trust" was applied by a *New York Times* reporter. As late as 1941 journalist Raymond Clapper included O'Connor in FDR's "unofficial cabinet,"

[6] Woodbury, Clarence. "The Man in the Middle." *The American Magazine*, September 1955, p. 103.

characterizing the members as "kindred spirits, articulate men, with hair trigger minds who spark the President's thinking."[7] Cognizant of their close partnership at Warm Springs, Tugwell characterized O'Connor as a brake on FDR's adventurous proclivities in their business ventures reflecting that, "Doc's present involvement was hard to describe. He was around a good deal; he sat in on strategy meetings, criticized, and did odd jobs when he was asked. Also he helped to keep things straight."[8] In *Working with Roosevelt* (1952), Samuel Rosenman reported that "Constant advice and assistance, and sometimes drafts of campaign speeches were furnished by Louis Howe and Basil O'Connor, who was a law partner, close friend and able and astute associate of the Governor. O'Connor worked on speeches very frequently with me, especially when Roosevelt was in New York City; he had a very effective and forceful manner of writing."[9]

Fishing Partners, 1931. New York Governor FDR, Basil O'Connor, and Almon Rasquin enjoy an afternoon of fishing at the "Snug Harbor" O'Connor home in Westhampton Beach, Long Island. Rasquin was Collector of Internal Revenue for the eastern district of New York. Roosevelt, in steel braces, holds onto the car and O'Connor's hand for support. Bettyann O'Connor appears in the background.

[7] FDRL, Basil O'Connor Papers. finding guide.
[8] Tugwell, Rexford. op. cit., p. xiii, 18.
[9] Rosenman, Samuel. *Working with Roosevelt*. New York: Harper & Brothers, 1952. p. 4.

The Democratic convention of 1932 proved a critical episode in the inter-active balance of the group, and the chronology of how the Roosevelt forces turned the tide of nominating votes toward FDR has been repeatedly exam-ined. At the Chicago convention, Louis Howe's suite in the Congress Hotel on Michigan Avenue was a kind of "command post" for the Roosevelt contingent, with O'Connor's suite adjacent. Tugwell and O'Connor labored feverishly on FDR's acceptance speech only to have much of their writing scrapped by FDR himself as he blended in Rosenman's version at the end of his unprecedented air journey to accept the nomination, dramatically capping the urgency of the moment. It was in this speech that Roosevelt first used the phrase "new deal": "I pledge you, I pledge myself, to a new deal for the American people." After their return to New York, Tugwell asked O'Connor why FDR did not resign as governor and direct his energies solely to the presidential campaign. O'Connor's explanation was simple: FDR was "poor." Or, rather, as Doc elaborated, FDR was "poor for a Roosevelt," understanding first-hand the implications of this unexpected revelation from managing the Warm Springs budget.[10] Not only were his organizational skills vital, O'Connor made personal contributions to the campaign from his own finances. During the campaign, some appealed to FDR to use his influence as governor to halt "short selling" on the New York Stock Exchange. O'Connor gave explicitly contrary advice, arguing that short selling (ie, selling a security that the seller has borrowed but does not own) was relatively unimportant; instead, he advocated the temporary closing of the exchange to forestall panic as had been done in 1914.[11]

Because he stood at the margins of the political limelight, O'Connor already had acquired the image of the "forgotten man" in the Roosevelt & O'Connor law partnership. FDR himself had used the phrase "forgotten man" in a radio address as a metaphor for the poor who needed economic assistance but were not getting it. The phrase was further popularized in the film *My Man Godfrey* in 1936. In fact, the *Boston Evening Transcript*, a month before the inauguration, depicted O'Connor in just this fashion, as "Roosevelt's Forgotten Law Partner," dropping a bit of hesitant speculation that he "may" be a member of a Demo-cratic "Brain Trust."[12] To forestall criticism of impropriety concerning FDR's interests outside the presidency, O'Connor officially eliminated Roosevelt's name from the partnership on July 10, 1933. FDR reluctantly agreed that this move was in their mutual interest but expressed his exasperation that he "hated the thought of terminating the fine old firm name of Roosevelt & O'Connor."[13]

[10] Tugwell, Rexford. op. cit., pp. 234, 247–49, 297.
[11] FDRL, Roosevelt Family Papers Donated by Children. Basil O'Connor, folder 7. June 21, 1932.
[12] Sanborn, Ralph. *Boston Evening Transcript*. February 1, 1933.
[13] FDRL, President's Personal File, 96. Basil O'Connor, 1938–1939. See also DART, Basil O'Connor Papers, Roosevelt and O'Connor law firm, bound volume; 1925–1933.

O'Connor reassured him, "You have been so perfectly fine through all these years, and I want you to know how deeply I appreciate it."[14] But if the dissolution of the law firm marked the end of an era in their relationship, O'Connor's service to the President during the tempestuous days of economic reconstruction continued unabated in tandem with their joint efforts to expand the fight against polio. When Carter Glass declined FDR's offer of a cabinet post as Secretary of the Treasury, FDR sought O'Connor's help to convince the Republican industrialist William Woodin to accept the position. Jean Edward Smith remarked that "Woodin, too, was reluctant, but Basil O'Connor convinced him to take the post during an hour-and-a-half cab ride circling through Central Park."[15] Much later, O'Connor ran interference for FDR again in attempting to bring the wayward campaign manager James Farley back into the Roosevelt fold in 1940. After the first inaugural in March 1933, O'Connor himself felt impelled to publish a personal tribute to FDR in a Dartmouth newspaper, "He makes people want to work with him and for him. He fills them with a desire to do things and do them well. He impels everybody to cooperate—a tremendous personality."[16] In return, FDR discussed nominating O'Connor to the US Supreme Court, a position the latter declined ostensibly because his own law practice was far more lucrative and he preferred to avoid the actual scrimmage of political life even though drawn to other prominent leadership roles. Earlier, New York Governor Herbert H. Lehman had sought O'Connor's appointment as Attorney General of New York with the same result.[17]

O'Connor's own impulse to cooperate was anchored deeply in the foundation of his personal friendship with Roosevelt, apart from the formalities of official roles and duties. This is evidenced quite clearly in the relationship between the two families. O'Connor's cordiality with Eleanor and the Roosevelt children matured slowly and incrementally, while FDR's connection to Elvira O'Connor seemed more immediate and personal. When Elvira traveled to Warm Springs to recuperate from an illness (not polio), Roosevelt wrote to her directly, "feel free to use my cottage," and Missy LeHand conspired with her afterward to reciprocate his generosity by furnishing it with a "hooked rug."[18] With O'Connor's daughters Bettyann and Sheelagh, Roosevelt was decidedly avuncular. Letters from FDR to the O'Connor girls were invariably signed "Uncle Franklin" or "Your Affectionate Uncle." When Sheelagh O'Connor expressed interest in starting a stamp collection,

---

[14] MDA, *The History of The National Foundation for Infantile Paralysis.* Volume II, Book 2. p. 77. See also DART, Basil O'Connor Papers, Roosevelt and O'Connor law firm, bound volume; 1925–1933.

[15] Smith, Jean Edward. *FDR.* New York: Random House, 2007. p. 293.

[16] MDA, ibid. Originally in *The Dartmouth.* May 3, 1933.

[17] MDA, Oral History Records. Stephen Ryan interview, June 13, 1983. See also Dartmouth College *Alumni Magazine,* December 1938.

[18] FDRL, President's Personal File, Box 96. Basil O'Connor, 1938–1939.

FDR's immediate advice to her, based on his own long experience with this hobby, was to "specialize!"[19] The two families enjoyed social visiting, but these occasions dwindled significantly during the years of the Roosevelt presidency. Invited to Bettyann's wedding at the O'Connor home in West-hampton on July 15, 1939, the Roosevelt family's tentative acceptance ended in regrets. Bettyann dutifully penned a letter to "Uncle Franklin" thanking him for the silver vegetable dishes with the note "wish you could be with us."[20] Even during the onerous years of the war, Roosevelt rarely failed to acknowledge birthdays, rites of passage, and anniversaries. He sent personal greetings to the Sidney Culver family (nee Bettyann O'Connor) on the birth of their first son in 1941 and congratulations to Elvira on her 25th wedding anniversary with Basil.[21] In those congratulations, FDR praised her husband to the point of portraying his own friendship with him as a kind of marriage. Elvira in turn expressed her incredulous delight that FDR could actually spare the time to write to her in his own hand (just as Allied forces were poised to invade southern Italy).[22]

O'Connor's standing with Roosevelt might easily have been compro-mised by the politics of his brother, John J. O'Connor, Republican chair of the Congressional Rules Committee whom FDR targeted for defeat in 1938.[23] In the muckraking column "Washington Merry-Go-Round," Drew Pearson labeled Basil O'Connor and his brother John "a worse pain in the neck" to FDR than any of his political enemies. Pearson portrayed Basil as having "his hand in lobbying deals diametrically opposed to the New Deal," and he accused John of obstruction of several beneficial pieces of legislation as the powerful chairman of the house rules committee.[24] But O'Connor was not loathe to report candidly to Roosevelt about allegations made in the press about the political machinations of the O'Connor brothers, as in a 1935 note that the "O'Connor boys had sold out to Associated Gas & Electric for $25,000."[25] If this family connection burdened O'Connor's allegiance to FDR with troublesome complexities from a scandal-mongering press, it failed to deflect the two men from their common purpose.

On one occasion, the two actually took a long vacation together. Roosevelt's affection for nautical affairs derived not only from his eight-year stint as Assistant Secretary of the Navy and his Florida houseboat travels on the *Larooco* as he recovered from polio, but from many seafaring experiences throughout his life.

---

[19] FDRL, President's Secretary File, Box 145. February 17, 1936.
[20] Ibid. July 6, 1939.
[21] Ibid. October 20, 1941.
[22] Ibid. August 28, 1938; October 27, 1943.
[23] Smith, Jean Edward. op. cit., pp. 413, 415.
[24] Pearson, Drew and Robert S. Allen. "The Daily Washington Merry-Go-Round," *Holyoke* (Mass.) *Transcript.* June 24, 1938.
[25] FDRL, Ibid. August 16, 1935.

In the summer of 1938, FDR and O'Connor traveled from San Francisco through the Panama Canal and on to Washington aboard the USS Houston, visiting the Galapagos Islands along the way. A 10,000-ton heavy cruiser, the Houston was one of Roosevelt's favorite ships of the US fleet. During World War II it became known as the "Galloping Ghost of the Java Coast" and was destroyed and sank on March 1, 1942 in the Battle of Sunda Strait in the Java Sea. En route to the east coast on this presidential junket, the Houston crossed the equator near the Galapagos Islands for a week's visit. The *New York Times* reported, "Roosevelt Sees 300 Houston 'Pollywogs' Changed to 'Shellbacks' by Father Neptune," a reference to a "crossing the line" ceremony, an obligatory rite of passage for sailors on their first time crossing the equator at sea.[26] After the ritual, sailors were designated "Sons of Neptune." With O'Connor, FDR enjoyed the ceremonial hijinks on the high seas and was photographed with Neptune, the Royal Sawbones, and the motley crew. On other days he enjoyed the opportunity for deep-sea fishing and added to the catch of sailfish for the Houston. After his return to Hyde Park weeks later, FDR reminisced about this fishing adventure with a glib witticism about Washington politics regarding "burrowing," a slang reference to a temporary political appointee who gains permanent civil service protection:

> *As a matter of fact, there had been so much discussion on previous trips, about the size and weight and length and species of fish, that this year I took a full-fledged scientist with me from the Smithsonian Institution in Washington, Dr. Waldo Schmitt, who was such a success that we decided to change the Smithsonian to "Schmittsonian." When we started from San Diego out on the West Coast, we ran down the coast to Lower California which, as you know, belongs to Mexico. In talking to Dr. Schmitt that first day, I said: "Is there any particular thing or animal that you would like to find?" He said: "Oh, yes, I am writing a monograph. I have been on it two years, and the one thing I am searching for in these waters of Mexico and the islands of the Pacific—I want to find a burrowing shrimp." "Well," I said, "Dr. Schmitt, why leave Washington? Washington is overrun with them. I know that after five years."[27]*

A month-long vacation in tropical waters was hardly Basil O'Connor's predilection, however, especially in light of his emergent role as president of The National Foundation for Infantile Paralysis (NFIP). On their return, Roosevelt telegrammed Elvira O'Connor that Doc was "a bigger and better man" thanks to this fishing trip disguised as a presidential retreat. In reply, Elvira confessed that "I really don't know of anything he's ever done that was so good for him."[28]

---

[26] *New York Times.* July 23, 24, 25, 26, 31, 1938.

[27] Rosenman, Samuel I. *The Public Papers and Addresses of Franklin D. Roosevelt.* New York: Random House, 1938. Informal remarks at a meeting of the Roosevelt Home Club, pp. 502–03. I am indebted to Cindy Pellegrini of the March of Dimes for explicating the political meaning of the term "burrowing."

[28] FDRL, President's Personal File, Box 96. Basil O'Connor, 1938–1939.

Crossing the Line, July 1938. FDR's affection for the sea and nautical affairs derived from his eight-year stint as Assistant Secretary of the Navy, his Florida houseboat travels in recovery from polio, and from seafaring experiences throughout his life. FDR is seen here aboard the USS Houston in the company of US Navy sailors participating in a Crossing the Line ceremony near the equator en route to the Galapagos Islands. Crossing the Line is an initiation rite that marks the first time a sailor crosses the equator at sea, after which he becomes a 'Son of Neptune.' The USS Houston, a 10,000-ton heavy cruiser, was one of FDR's favorite ships of the US fleet, known during World War II as the "Galloping Ghost of the Java Coast." The ship was destroyed and sank in the Battle of Sunda Strait in the Java Sea on March 1, 1942. This photo is from a private collection of Basil O'Connor in the March of Dimes Archives.

Basil O'Connor was a facilitator, fixer, factotum, and master fund-raiser, but first of all he was a shrewd attorney whose métier was international in reach and whose services to Roosevelt were varied in the extreme. His value to Roosevelt "accelerated uniformly" without letup over the course of FDR's presidency, culminating in high-profile positions as the head of the NFIP and American Red Cross. The range of his talents was prodigious, and he brought to the table advice on the political import of corporate affairs, personal recommendations on ambassadorial and New Deal agency appointments, and legal analyses of Roosevelt's personal estate, insurance, and taxes. With the outbreak of war his advice to Roosevelt ranged over a host of strategic issues such as navigation, shipping, trade routes during wartime, war-risk insurance, hoarding, stock values, and oil reserves. In 1939, O'Connor became one of the first trustees of the Franklin D. Roosevelt Library, Inc., leading the efforts to raise money and erect a building to house the president's library and vast collection of papers and memorabilia. This was the first presidential library in the system now administered by the National Archives and Records Administration (NARA) of the United States. After the death of Sara Roosevelt, FDR's mother, O'Connor took charge of the appraisal and sale of the Roosevelt family town house at 47–49 East 65th Street in Manhattan (where FDR began his recovery from polio in October 1921), and when FDR's trusted secretary Missy LeHand suffered a stroke in 1941, FDR asked O'Connor to revise his will, with plans to bequeath to Missy half of his estate. O'Connor was adamantly opposed to an alteration that would remove the Roosevelt children as beneficiaries from the will, and he said so, forcefully. An argument ensued between them, but in the end O'Connor demurred, and the President got what he wanted. Afterward, O'Connor, FDR's son James Roosevelt, and Henry T. Hackett, an attorney in Poughkeepsie, New York became the executors and trustees of the FDR estate.[29]

The "Fifth Wheel," 1933. Basil O'Connor's hobby of collecting signed portraits of friends and business colleagues reaches a culmination in this composite photograph of President FDR and his three closest administrators of daily business in the White House. The inscriptions read: "Dear Doc, Pack up your troubles in your old kit bag" (Louis McHenry Howe, FDR's political advisor); "with all best wishes," (Stephen Early, White House Press Secretary); "From one tailor-made man to another" (Marvin H. McIntyre, Presidential Secretary). At the bottom, FDR's inscription to his friend Doc reads: "For D. B. O'C. Fifth Wheel from the other four wheels."

[29] FDRL, President's Secretary File, Basil O'Connor, 1942–1944. See also Goodwin, Doris Kearns. *No Ordinary Time*. New York: Simon & Schuster, 1994. p. 246.

From the moment he became Governor of New York, O'Connor savored political cartoons about FDR. He clipped and collected these newspaper cartoons religiously through the four terms of Roosevelt's presidency, always curious about how "the boss" was praised by his admirers or pilloried by his detractors in order to gauge the public mood. Like Roosevelt, O'Connor was indeed a collector, and his other notable collection was a portrait gallery of signed photographs of Dartmouth alumni, colleagues at the bar, and professional men from business and academia that he had met over the years. He began to build this collection around the time Roosevelt became governor and initially referred to it as his "rogue's gallery of those who have been close to the governor." He displayed the photos with admiration and pride in his law office.[30] He never hesitated to request an autographed portrait, and his "Gallery of Distinguished Gentlemen" included photographs of the leaders of Warm Springs and the Brain Trust, New Deal appointees, and politicos of every stripe. American lawyer and statesman Elihu Root, whom O'Connor much admired, responded to a photo request in 1934: "I am ashamed to give anyone the last photograph taken of me, which is now 10 years ago. It is sailing under false pretenses." Another poked fun at O'Connor: "I am flattered by your request for a signed photograph but the first question is whether this is for exhibition purposes (where, of course, it should be) or for the reserve file. Just how often do you change them on the walls; is the arrangement copyrighted; do you make the arrangement with an eye to the probability of unfair competition, libel by reason of undue proximity to someone less handsome, and who makes the selection?" Dr. Michael Hoke needled him, "I feel very much flattered by your request for my photograph. You must be going to chase the rats away."[31] Hoke's political wit was not lost on O'Connor, and his own witty banter with FDR is a private measure of their close camaraderie. On one occasion he teased Roosevelt about how the press seemed to exaggerate his influence over both the President and the Secretary of State: "Just in case you should read how successfully I influenced you and Secretary Hull to make a $30,000,000 loan to China—I really didn't! It's just another retainer I never received!" O'Connor had been singled out as one of four attorneys that had pressured Roosevelt and Hull to approve a $30 million loan to China that resulted in little more than a briefly newsworthy law suit. O'Connor's affection was constant, and yet his importunate pleas to FDR could be downright shrill: "*Please, please, do not* send [my speech] to your hackers to perform an OOPHORECTOMY on it. I just couldn't stand that!" His demands to

---

[30] FDRL, Rexford Tugwell Papers, O'Connor to Tugwell, December 27, 1932.

[31] DART, Photography Collection of Basil O'Connor. Four boxes of photo portraits. NYSA, Basil O'Connor Records. Michael Hoke to O'Connor. April 23, 1934.

Missy LeHand or to Grace Tully to squeeze in a last-minute appointment with the President sometimes bristled with impatience: "Tell him it's about *nothing*. I've given up trying to solve problems. There are too many of them!"[32] Yet on one of the darkest days of the war in 1942, with the Allies beleaguered on all fronts and the nation consumed by the constant shock of dismal news, O'Connor felt free to vent his bile about the war rationing policies of Secretary of the Interior Harold Ickes and so tossed off this limerick in a letter to his dear friend in the White House:

> There was a lady of fashion
> Who had a terrific passion;
> As she jumped into bed
> She casually said,
> "Here's one thing that Ickes can't ration."[33]

This surely brought forth a generous chuckle from the President. William Hassett, who became FDR's correspondence secretary in 1944, considered Basil O'Connor to be one of FDR's "closest friends and business associates."[34] A lasting indication of this is evidenced in one of the most treasured portraits in O'Connor's "rogue's gallery." Dated October 7, 1933 it features the composite portraits of President Roosevelt, Louis Howe, Stephen Early, and Marvin McIntyre in a single panel—four men at the supreme echelon of power in the White House. O'Connor had actually plagued Missy with requests to help him acquire this photo. Inscribed to "Doc" and autographed by each of the four, FDR's culminating inscription to his friend Daniel Basil O'Connor includes him without question in their intimate company: "*To D.B.O'C., fifth wheel, from the other four wheels.*"[35]

Basil O'Connor was the critical "fifth wheel" among Roosevelt's daily advisors, the one with the most staying power and breadth of latitude well beyond any of the limited configurations of the Brain Trust, however one might enumerate their company. O'Connor's assistance to FDR as an unofficial "advisor without portfolio" presaged the role he would assume when the NFIP was established, for despite the ostensibly nonpartisan aims of the new foundation as an autonomous domestic organization whose mission was to defeat polio, its early maneuvering for local support across the nation

[32] FDRL, President's Secretary File. Basil O'Connor to Roosevelt; February 8, 1940; January 24, 1944; O'Connor to Grace Tully; July 22, 1943.
[33] Ibid. April 16, 1942.
[34] Hassett, William D. *Off the Record with F.D.R., 1942–1945*. London: Allen & Unwin, 1960. p. 33.
[35] MDA, Photography Collection. Autographed portraits, 1933.

and globally required knowledge of local issues as well as the diplomatic finesse that a person like O'Connor might provide. Together, FDR and O'Connor positioned the NFIP in its earliest years as an American institution aligned with the New Deal that functioned in the international arena with a role in diplomacy and aid beyond the specific focus on polio. Even as O'Connor was building an infrastructure of volunteers and creating his medical committees he was helping in some cases to guide the NFIP as an instrument of projecting American power overseas as needed by President Roosevelt. His assignments were not as romantically dashing as, say, those of Harry Hopkins in his wartime flights to Great Britain and the Soviet Union to help secure a wartime alignment of the Allies. Yet from the founding of the NFIP in 1938 to the conclusion of his term with the American Red Cross in 1949, O'Connor used his positions to introject elements of Roosevelt's Good Neighbor policy in Latin America and as far afield as the Philippine Islands, even after FDR's death in 1945.

One stark example was Argentina. In 1943, the Argentine Department of Health solicited aid from the NFIP to help combat a polio epidemic, and after O'Connor had consulted with Roosevelt, the Foundation dispatched Mary Kenny (Sister Elizabeth Kenny's daughter) and Rutherford John, an orthopedist from Philadelphia, to Buenos Aires to provide technical support. Argentina had not severed ties with either Nazi Germany or fascist Italy, and after a military coup during the polio team's visit, the NFIP reconsidered the initial plan and withdrew its support entirely. This precipitated the anger of Sister Kenny herself, leading to a falling out with O'Connor and the NFIP, whose relationship was already argumentative. Years later, in 1956, the NFIP would again intervene in Argentina at the request of the US State Department. The citizens of Buenos Aires were frantic: schools were closed, parents held day-long vigils for paralyzed children at hospitals, and many people scrubbed sidewalks and streets or whitewashed porches and buildings in desperation. Besides the tangible medical aid and advice provided by the NFIP, both the State Department and O'Connor considered it a unique opportunity for the United States to improve relations between the two countries. But there were earlier situations that were complicated by the existence of Birthday Ball Committees and certified chapters of the NFIP that existed in far-flung extraterritorial settings that thrust the NFIP into the international arena.[36]

[36] Rogers, Naomi. *Polio Wars: Sister Kenny and the Golden Age of American Medicine.* New York: Oxford University Press, 2014. pp. 110–14.

The chapter system of the NFIP devolved logically and practically from the local organization of president's birthday balls by groups of citizens in the 1930s. Beginning in 1939, with the creation of the first NFIP chapter in Coshocton, Ohio, any group of volunteers in any community could affiliate with the NFIP by creating a county chapter. The rules were simple, and certification by NFIP headquarters in New York was a must. A group of citizens, usually consisting of a community business leader, physician or public health official, and/or any combination of housewives, mechanics, bankers, journalists, or persons concerned, could apply to the NFIP to become certified as an official chapter. The popularity of the March of Dimes campaigns (see Chapter 5: The March of Dimes in World War II) boosted the formation of these local units enormously in the early 1940s; within 2 years, by February 1942, there were 2,232 county chapters, and within a decade there were over 3,100 chapters, all involved in fund-raising, epidemic relief, polio education, and patient aid. But besides the evident fertility of NFIP chapters springing up across the nation, local chapters were also created far beyond the national boundaries of the United States in extraterritorial annexations and protectorates, namely in Alaska and Hawaii (not to become states until 1959) as well as in the Panama Canal Zone, the Philippine Islands, and Puerto Rico. March of Dimes activity also occurred on Guam and Okinawa, and chapters were proposed for these islands but never materialized. The history of these extraterritorial chapters provides a glimpse of O'Connor's involvement in international affairs as an element of his ongoing and unofficial advisory role in the Roosevelt administration. While local chapters in the continental US retained a level of local autonomy though coordinated through the NFIP headquarters in New York, it was O'Connor himself who oversaw the particular situations of March of Dimes activities on foreign soil and remained available to dictate or modify policy when the need arose. At that time (1938–44), his operational flexibility lay still within the orbit of FDR's closest advisors who became semiautonomous independent agents on ad hoc missions.

In Alaska, a Committee for the Celebration of the President's Birthday operated for 3 years to 1941, when 18 people held a birthday ball at Mt. McKinley National Park. Two small NFIP chapters had been established, but Peter J. A. Cusack (NFIP Executive Secretary) advised the creation of a chapter covering the entire territory of Alaska to "absorb the two small Chapters now in existence."[37] O'Connor contacted the Governor of the

[37] MDA, Chapter Administration Records. Alaska, 1939–1950. June 11, 1941.

Territory of Alaska in Juneau, Ernest Gruening, to take on the position of NFIP "state" chairperson to form a single territory chapter to absorb the two already created. O'Connor gave his standard pitch to Gruening: "The primary purpose of the Territorial Chapter would be to render financial aid to victims of infantile paralysis who cannot pay for the necessary treatment. It would also assist the community in the event of an epidemic and carry out the Foundation's educational program in its area." But with the advent of war in the Pacific theater Gruening balked. He replied to O'Connor on June 5, just after Japanese forces had attacked Dutch Harbor on Amaknak Island in the Aleutian chain and pleaded that "almost all energies in the Territory at this time are being expended in the war effort." He noted that polio was almost nonexistent in Alaska, a total of only 30 cases from 1933 to 1947. This changed slightly after the war when Gruening became a chapter chairperson, and O'Connor maintained personal contact with him through the mid-1950s since his jurisdiction was extraterritorial, requiring the personal administration of the NFIP head.[38]

A similar state of affairs held in Hawaii, but in this case the war was *the* dominant issue as it would be in the Philippine Islands, Okinawa, and Guam. The NFIP chapter of the County of Hawaii created in 1939 tended to a few mild cases of polio until the attack on Pearl Harbor brought the United States into the war, reorienting all activities to a permanent war footing to respond to the emergency. By 1943, the chapter had resumed a semblance of normal functioning. B. J. McMorrow, Chapter Chairperson and Acting Health Officer of Hawaii, told O'Connor, "Due to the tenseness and unusual conditions existing on this Island in the months following the beginning of the war, no effort was made to conduct the affairs of the Hawaii Island Chapter of the National Foundation in a routine and normal manner. ... No organized fund-raising campaigns have been carried on since 1941. The Chapter, however, has not been inactive as it has paid for the transportation to Honolulu for treatment of children crippled through infantile paralysis and has distributed educational literature."[39] O'Connor applauded McMorrow's efforts in resuscitating the chapter in the wake of Pearl Harbor and regretted the wartime transfer of most of its officer personnel. He reminded McMorrow that the reorganized chapter should remain "nonsectarian and nonpolitical."[40] By 1945, there were two chapters,

---

[38] MDA, Chapter Administration Records. Alaska, 1939–1950. May 29, 1942; June 5, 1942.
[39] Ibid. Hawaii, 1939–1962. June 3, 1943.
[40] Ibid. June 8, 1943; September 5, 1943; October 11, 1944.

the Hilo chapter on Hawaii and the Honolulu chapter on Oahu. After the war, NFIP territorial representative Carolyn Patterson (nee Kingdon) coordinated NFIP activities across the broad stretch of the Pacific from her post in Honolulu. She turned down a request to develop a chapter on Okinawa in 1954 "since the Ryukyu Islands are considered foreign territory and the United States has jurisdiction only insofar as administration is concerned." A "March of Music" program had raised almost $25,000, mainly from US staff stationed there after the war. A similar situation obtained on Guam, which was then classified as an "unincorporated territory," ceded to the United States by Spain in 1898. Kingdon reported that the Guamanians were largely in favor of a local chapter of the NFIP but strictly to keep hold of the March of Dimes money contributed by armed forces personnel who had raised $38,000 in the 1953 March of Dimes. In any event, US military personnel and citizens based on Guam, Hawaii, and Okinawa had the same status with regard to polio care financed by the NFIP. On occasion, the NFIP paid for polio care even on Guam.[41]

The situation in the Commonwealth of the Philippines was entirely different, for an active unit had been created in Manila on May 31, 1939, a year after Roosevelt founded the NFIP. By Administrative Order No. 96 issued by Manuel L. Quezon, President of the Philippines, styled "Creating President Roosevelt's Poliomyelitis Committee," it was acknowledged that the birthday balls for President Roosevelt "to control and combat infantile paralysis" had been held annually since 1935, and that the committee would thereafter manage the funds collected in these balls for the "National Committee" in the United States. The order stipulated that 50% of the funds would be retained locally in keeping with NFIP practice. However, the work of the committee was conflated with another mission: tuberculosis control. President Quezon had suffered from tuberculosis from the 1920s; and after the Philippines fell to the Japanese in 1941, he formed a government in exile in the United States and ultimately died in the tuberculosis cure cottages of Saranac Lake, NY in 1944. Just as O'Connor would refuse to compromise NFIP policy on federated fund-raising later in the 1940s, mixing the purposes of these two health problems in a single fund-raising effort seemed bound to cause difficulties. When Keith Morgan heard about the Philippine committee, he immediately wired O'Connor, anticipating his disapproval. Inevitably, O'Connor tried to turn the situation around by sending the Filipinos application materials to organize a proper NFIP chapter since the committee "has never been formally chartered by this

[41] MDA, ibid. July 23, 1954; September 23, 1952.

foundation." However, the American community in the Philippines who might engage in March of Dimes activities was comparatively small, and both O'Connor and Morgan realized the cooperation of the two governments over this issue and others was vitally important in the developing situation of the Pacific, especially since the commonwealth government also celebrated President Quezon's birthday to raise funds to fight tuberculosis.[42]

Francis B. Sayre, US High Commissioner to the Philippine Islands, then wired O'Connor for advice. He admitted that though American participation in President Quezon's celebration for tuberculosis was "none too great," the NFIP should accept the cross-participation of Americans and Filipinos in each other's events. He cautioned, however, that pressing for the organization of NFIP chapter "may not be desirable." The problems of distance, lack of equipment, and minimal funds militated against insisting on the creation of a formal chapter. Although O'Connor still preferred an NFIP chapter he admitted the situation was awkward because FDR permitted the use of his birthday *in return for funds raised for NFIP* [emphasis added].[43] Sayre's radio address of January 28, 1940 honored Roosevelt, hailing as a man who "in spite of the handicap of a shattering disease, refused to surrender to disaster."[44] But a larger disaster was just on the horizon: the Philippine Islands fell to the Japanese invasion that brought America into the war. Although the NFIP chapter of the Philippines was organized on March 18, 1941, it collapsed and remained completely inoperative during the course of the war.[45]

The Panama Canal Zone presented a less momentous problem, one of financial management. The Canal Zone was an unorganized territory, controlled by the United States from 1903 to 1979. American citizens residing there organized birthday balls from the mid-1930s, and a movement to form an NFIP chapter stirred in 1941 with the advent of the March of Dimes. Yet the Canal Zone Committee for the Celebration of the President's Birthday asked the NFIP to exempt the zone from consideration as a potential chapter because it was "strictly a military-government reservation." Richard D. Moore, Executive Chairman of the committee met with O'Connor, Morgan, and D. Walker Wear soon after the Pearl Harbor attack

---

[42] MDA, Chapter Administration Records. Philippine Islands, 1939–1962. July 19, 1939; September 15, 1939; December 13, 1939.

[43] Ibid. January 23 & 24, 1940; February 29, 1940.

[44] Ibid. Report of the Committee for the Celebration in the Philippines of President Roosevelt's Birthday, 1940.

[45] Ibid. January 8, 1941; January 5, 1950.

to present his position, and O'Connor agreed to block any formal plans to create a chapter due to "national defense problems faced at the time."[46] But neither the war nor his decision prevented Americans living in the Canal Zone to curtail their birthday ball hoopla or to cease sending revenue to the NFIP. After the celebration in 1942, Glen E. Edgerton, Governor of the Panama Canal Zone, wrote to Keith Morgan to suggest that local monies be placed in trust as a contingency fund. Edgerton kept to his word and remitted $7,000 to be held in safekeeping by the NFIP, and O'Connor created "The NFIP, Inc. Panama Canal Zone Account," a special trust fund at the Bank of New York.[47] The Canal Zone revenue was not inconsiderable: $12,101 in 1942 and $35,469 in 1943, a sum then invested in war bonds. By 1948, local interest in forming a Canal Zone chapter had not died down but since the American community consisted primarily of government employees and their families, the NFIP preferred to keep to an unofficial committee since it was already "accomplishing what they are supposed to do," that is, to raise money each year. From 1939 to 1950, the committee raised $222,000 for the March of Dimes. In 1955, the NFIP ensured that the Salk polio vaccine was made available to all children residing in the Panama Canal Zone.[48]

By contrast with Panama, the territories of Alaska, Hawaii, and Puerto Rico all had organized NFIP chapters. The Puerto Rico chapter was the most intractable of all. In 1939, Roosevelt appointed Admiral William D. Leahy as Governor of the Territory of Puerto Rico as an administrator of the Puerto Rico Reconstruction Administration, succeeding Leahy's two-year stint as Chief of Naval Operations from 1937. Leahy was sent in to replace Governor Blanton Winship, whose repressive policies led to widespread social unrest. Leahy's assignment was short lived, a mere 15 months, as FDR redirected the admiral's diplomatic expertise toward a more volatile situation as the US Ambassador to France during the formation of the Vichy government in an attempt to subvert its collaboration with Nazi Germany. Even that proved a relatively brief assignment, as Roosevelt soon recalled Leahy to serve as head of the Joint Chiefs of Staff to help FDR "run the war." In Puerto Rico, however, Leahy helped to establish American military bases on the island and initiated several public works projects. In keeping with the ostensible American move toward the

[46] MDA, Chapter Administration Records. Panama Canal Zone, 1941–1961. December 24 and 30, 1941; October 18, 1948.
[47] Ibid. February 25, 1942; April 21 and 27, 1942.
[48] Ibid. May 26, 1943; December 23, 1948; May 22, 1950.

benevolence of local development, Admiral Leahy's wife, the former Louise Tennant Harrington, was recognized as the "Honorary President" of the local chapter of the NFIP for the Territory of Puerto Rico. This was a safe but fairly prominent way to motivate public opinion in the direction of the charitable tendencies of Rooseveltian statecraft channeled through the influence of Basil O'Connor and the NFIP.[49] President Roosevelt appointed Rexford Tugwell to be Governor of Puerto Rico in 1941. As the last appointed American governor he served for the duration of the war, avidly supporting Puerto Rican self-government despite local calls for his removal in 1942. Tugwell, one of the key architects of Roosevelt's New Deal, was no stranger to controversy, and his relationship with O'Connor had remained intact from their formative experiences at the core of the Brain Trust a decade earlier.

Soon after the appointment of Admiral Leahy, Basil O'Connor and Keith Morgan broached the topic of establishing an NFIP chapter in Puerto Rico with Ana Maria Valdes de Iriarte, the wife of Senator Celestino Iriarte Miro, a ranking member of the Senate of Puerto Rico associated with the Republican Union party and the Alianza Puertorriquena, a unity party that had dissolved in 1932. Mrs. Iriarte became "City Chairman" of the new chapter, and she led a marginally successful March of Dimes campaign in 1940 in San Juan. Her credibility with the NFIP was not to last very long, however. Mrs. Iriarte was a Republic Union partisan who loved to play at politics.[50] Once the Puerto Rico chapter had reserved a fund for polio cases, O'Connor approved transfer of $4,000 from the chapter to the Commissioner of Health to maintain four beds in the Convalescent Home for Crippled Children for one year under the direction of George L. Kraft, MD of the Department of Health.[51] Mrs. Iriarte redirected the funds to a Presbyterian Hospital Medical facility known as Clinica Pereira Leal where her husband Senator Iriarte was Director with a substantial financial interest. Polio cases fell under the care of Peter E. Sabatelle, MD, an orthopedic surgeon and prominent member of the NFIP chapter, and Leon Sheplen, MD, who had been trained in convalescent care at Warm Springs. NFIP Director of Chapters Peter Stone approved the request, but came to question it when he learned that the four beds were for general cases, despite a waiting list of

[49] MDA, Chapter Organization Certificates Records, Puerto Rico Chapter.
[50] Ibid. The chapter application is dated April 18, 1940; the certificate of organization is May 20, 1940. Chapter personnel include: Honorary President, Mrs. W. D. Leahy; Chairman, Mrs. Celestino Iriarte (*Asociacion Bibliotecaria de Puerto Rico*), et al.
[51] Ibid, 1940–1955. October 18 and 20, 1940.

350 polio cases. Stone demanded a complete reevaluation of the case which rankled Mrs. Iriarte, who protested to the chapter committee, citing the many contracts that Clinica Pereira Leal had with the Work Projects Administration (WPA), Civilian Conservation Corps (CCC), Federal Land Bank, Farm Security Administration, Naval Base, Teachers Association, Union Puertorriquena, and National City Bank.[52] Dr. Sheplen complained of her "Machiavellian machinations" and "sleight of hand." He admitted to Stone, "I have never seen such gangster politics," as Mrs Iriarte overruled the committee's recommendation to ask NFIP for a final decision. As a result, the chapter nearly dissolved even as Dr. Sheplen concluded in a formal study that polio was "an endemic disease of proportionally large incidence in Puerto Rico" with 341 cases under 2 years of age in the past two years.[53]

The chapter became mired in the dissension of competing views over the affair, and Governor Tugwell intervened. He requested the Puerto Rico Commissioner of Education, Jose M. Gallardo, to take charge of the birthday ball campaign in 1944 since Mrs. Iriarte was away on an extended visit to the United States. When Gallardo attempted to reorganize the chapter, Mrs. Iriarte returned and called her own meeting after the 1944 birthday ball, fired the treasurer, and replaced her with her own choice. She objected to Gallardo's choice of secretary because of her political affiliations, stating with temerity, "I feel that we should not mix party labels with the eradication of infantile paralysis." Gallardo, Tugwell, and O'Connor conferred over the best course of action. O'Connor appointed Tugwell himself to head the annual March of Dimes campaign, but with no authority over the chapter. The contretemps fizzled out in the petty pandemonium of legal action brought by Mrs. Iriarte over the control of chapter records, and though she resumed her position as head of the chapter, its accomplishments paled in the face of the energy she expended in her personal enmity toward Tugwell. Dr. Sheplen reported in 1946 "that the National Foundation chapter … is and has been ineffective in providing medical care for polio cases and uncooperative with the Crippled Children's Commission," rendering care to only five patients in the past five years. He blamed "local politics for much of the difficulty."[54]

---

[52] MDA, Chapter Administration Records. Puerto Rico, 1940–1955. July 28, 1940; August 15 and 28, 1940.

[53] Ibid. "Infantile Paralysis in Puerto Rico" by Leon B. Sheplen, MD and Blanca H. Trelles, MD.

[54] Ibid. June 7, 1944; July 17, 1944; September 19, 1946.

Basil O'Connor and Carlos Finley Institute, 1943. The overwritten caption reads: "Intimate Dinner in honor of His Excellency Dr. Manero, Secretary of Public Health of Mexico and Hon. Basil O'Connor, pres. of The Carlos Finlay Institute and pres. of Infantile Paralysis Foundation/Given by Dr. R. Castroviejo and Dr. J. Cervantes./N.Y.A.C./April 21, 1943." The Carlos Finlay Institute, established in 1941 to foster research and scientific development among Caribbean nations, named O'Connor as President of its Executive Council. Cuban President Fulgencio Batista installed O'Connor, Morris Fishbein, and other physicians of an American delegation to Cuba as Commanders of the Order of Finlay in a ceremony in Havana in January, 1942.

Despite its rocky beginning, the Puerto Rico chapter turned around to become the only extraterritorial chapter to continue into the 21st century (Alaska and Hawaii eventually developed state chapters). Basil O'Connor's international activities were propelled far beyond his management of the NFIP with his appointment to lead the American Red Cross in 1944 (see Chapter 7: Basil O'Connor and the American Red Cross). He continued to train his attention on the medical and public health politics of Latin America, particularly Mexico and Cuba, even as his post at the Red Cross redirected his international activity toward Europe in the massive war relief programs beginning in 1944 and 1945. In 1942, the Finlay Institute of the Americas

(Carlos Finlay Institute) in Havana, Cuba named him President of its Executive Council. He and his friend and advisor Morris Fishbein traveled to Havana in January 1942 in an American delegation for a ceremony to launch the Institute. He and his four American associates were installed as Commanders of the Order of Finlay by Cuban President Fulgencio Batista, and O'Connor had a private consultation with Batista afterward, bringing FDR's warm greetings of friendship. The Institute was created to foster science research and education in the Caribbean region, in honor of Carlos Finlay, a Cuba physician who first theorized that the mosquito is the disease vector for yellow fever.[55]

The NFIP was a domestic organization with a broad international outlook, as is the contemporary March of Dimes. During the polio era the Foundation was unsparing when calls for assistance arrived from neighboring countries and overseas. The first of these came from Canada in 1941, and Medical Director Donald Gudakunst traveled to Winnipeg to provide advice on a polio epidemic then raging in the province of Manitoba. In the aftermath of World War II, even as polio epidemics in the United States consumed the financial resources of the NFIP, it responded to polio emergencies in Belgium (1945), Germany (1947), Iceland (1949), Mexico (1951), Argentina (1956), the Marshall Islands (1958), and Japan (1960) among other nations of Europe, South America, Asia, and Africa. The NFIP arranged the shipment of iron lungs and medical equipment, dispatched teams of epidemiological investigators and medical advisors, and offered supplies of gamma globulin and, by the late 1950s, polio vaccine. By the time the Salk vaccine was declared effective in 1955, the NFIP was seen as the world leader in the fight against polio. After the mission change to birth defects prevention the Foundation assisted in the development of the International Clearinghouse for Birth Defects Surveillance and Research at the First Working Conference for Birth Defects Monitoring in Helsinki, Finland and was instrumental in founding the World Alliance of Organizations for the Prevention of Birth Defects in 1994 with the assistance of several European health organizations. The March of Dimes Office of Global Programs was created in 1998. All of these are elements of the legacy of Basil O'Connor and the internationalism of his perspective within the Roosevelt administration beginning with his role as a member of Roosevelt's Brain Trust.

---

[55] DART, Basil O'Connor Papers. General Correspondence, news clippings.

CHAPTER 5

The March of Dimes in World War II

Dr. John Paul, founding member of the Yale University Poliomyelitis Study Unit and first research grantee of the National Foundation for Infantile Paralysis (NFIP), famously compared the creation of the Foundation to the "sudden appearance of a fairy godmother of quite mammoth proportions who thrived on publicity" in his *History of Poliomyelitis*.[1] The perception of the new polio foundation as a kind of benevolent behemoth calls attention to its meteoric emergence in the new era of global communications media, but it was the power of the presidency that was the critical factor in its early successes. Franklin D. Roosevelt (FDR) had mastered the art of reaching radio listeners via regular "fireside chats" with a sense of intimacy heretofore unheard of, and his prominence as a heroic leader was unparalleled. Within its first 8 years, from its founding in 1938 to the final year of World War II, the NFIP amassed an impressive body of accomplishments toward the ultimate goal of ridding the country of polio. Its trustees were comprised of the top business leaders in the nation, it established a nationwide system of local chapters from the stalwart volunteers of the Birthday Ball Committees, and it methodically began to finance the medical research required to uncover the mechanisms of disease. In these years, the NFIP rushed emergency aid to polio-affected cities from Tacoma, Washington to Miami, Florida; established a depot for therapeutic splints in Austin, Texas; opened a polio hospital for African-Americans in Tuskegee, Alabama; recruited nurses from the American Red Cross for epidemic service; initiated epidemic preparedness meetings in Columbus, Ohio; and appropriated $1.2 million to train physical therapists desperately needed to treat the growing numbers of the polio disabled.[2] It had rushed emergency aid to Mississippi in August 1941 to support hospitals overwhelmed with polio cases; this occurred at a time when hospitals refused to accept such cases until the period of contagion (21 days) had passed. It issued a $3,000 emergency grant for epidemiological

---

[1] Paul, John R. *A History of Poliomyelitis*. New Haven: Yale University Press, 1971. p. 311.
[2] MDA, History of The National Foundation for Infantile Paralysis Records, Series 2: Typescripts. Highlights map, 1941.

Friends and Partners
ISBN 978-0-12-803597-9
http://dx.doi.org/10.1016/B978-0-12-803597-9.00005-9

field studies in Jasper, Alabama to Drs. John Paul and James Trask who claimed, "This is the first study which to our knowledge has been done in which every individual within a given epidemic area has been tested."[3] From its first medical conference in 1940 to the boisterous bandwagon of the annual March of Dimes campaigns, the Foundation forged ahead with rapid strides. The summer of 1942 found NFIP President Basil O'Connor recovering from abdominal surgery at the Columbia Presbyterian Medical Center in New York City, but the infrastructure of fund-raising, research grants, and patient aid in the NFIP had by then been successfully established. O'Connor recovered and was back at his desk at 120 Broadway in time for the 1943 March of Dimes. His determination was unshakable, and his modus operandi was captured succinctly in the framed placard on his office wall with his favorite slogan "*What Are the Facts?*" But how had he and Roosevelt brought the polio fight to this wider success?

In December 1934, the President's Birthday Ball Commission for Infantile Paralysis Research had been created as an independent organization that disbursed funds raised by the Birthday Ball Committee for polio research. Paul de Kruif—a microbiologist who had popularized the romance of medical research with books like *The Microbe Hunters* (1926)—was Secretary of the Commission. In its 1937 progress report for the Commission, de Kruif indicated that though its funds had been fully allocated to polio researchers at several universities, the dire need for a permanent source of funding remained strong if the Commission was to coordinate its several lines of research effectively. In their efforts to make America "polio conscious," Roosevelt and O'Connor had capitalized on the nationwide popularity of the annual birthday balls to expand services and patient aid at Warm Springs, and next they began to discuss how best to channel the momentum of the Commission's work. It was obvious to them that neither Georgia Warm Springs Foundation (GWSF) nor the Commission could fulfill the role of a national organization even though national attention had been focused on its polio mission for four years. Keith Morgan presented to O'Connor a financial plan he called "X = $7,000,000" envisaging that $7 million would be sufficient to fund polio research for a 10-year period. At first, a new agency was not considered, but by July 1937, their planning had turned toward the creation of a new foundation with O'Connor at its head. Originally to be named the "Roosevelt Foundation for Infantile Paralysis" and connected to the Birthday Ball Committee as its fund-raising arm,

---

[3] MDA, Committee on Research—Basic Science Grants Records. Mississippi and Alabama, 1941.

FDR and O'Connor decided to forego using the name "Roosevelt" so explicitly to distance the President from its management.[4] O'Connor, Morgan, and FDR all sought to avoid any further charge of partisan politics relating to the GWSF in a national effort that would supersede the institutional capabilities of Warm Springs. In truth, FDR was distanced from the NFIP only by the onus of presidential responsibility; his personal interest in the NFIP and its *March of Dimes* never flagged. Thus, on September 23, 1937 he issued a statement creating the NFIP; in part, he said:

> My own personal experience in the work that we have been doing at the Georgia Warm Springs Foundation for over ten years, lead me to the very definite conclusion that the best results in attempting to eradicate this disease cannot be secured by approaching the problem through any single one of its aspects, whether that be preventive studies in the laboratory, emergency work during epidemics, or after treatment. For over ten years at the Foundation at Warm Springs, Georgia, we have devoted our effort almost entirely to the study of improved treatment in the aftereffects of the illness. During these years other agencies, which we have from time to time assisted, have devoted their energies to other phases of the fight. I firmly believe that the time has now arrived when the whole attack on this plague should be led and directed, though not controlled, by one national body. And it is for this purpose that a new national foundation for infantile paralysis is being created.[5]

In a veiled reference to his own experience, Roosevelt went on to explain that the new foundation would "endeavor to eliminate much of the needless aftereffect of this disease—wreckage caused by the failure to make early and accurate diagnosis of its presence." He underscored the fact that while paralysis often leads to powerlessness and dependence, the renewed fight would emphasize maximizing the *independence* of the polio disabled in keeping with the high ideals of Warm Springs. He cautioned that the "public has little conception of the patience and time and expense necessary to accomplish" the restoration of this independence in the recovery from polio, as he knew all too well.

The title of the new foundation, "The National Foundation for Infantile Paralysis," was first used days later on September 29 in a memo to Basil O'Connor from William F. Snyder, an attorney at O'Connor & Farber, asking his concurrence in drafting a Certificate of Incorporation using this name to be filed with the New York Secretary of State. O'Connor gave the go ahead. The name was first used publically in a press release of November 25, 1937, which announced its Board of Trustees, many of whom were

---

[4] MDA, History of The National Foundation for Infantile Paralysis Records. Volume II, "Raising Funds to Fight Infantile Paralysis." Book 2, pp. 209–15.

[5] FDRL, FDR speech files. Box 34, NFIP annual report, 1939.

members of Roosevelt's New Deal agencies, such as George E. Allen of the Federal Deposit Insurance Corporation (FDIC), or those, like Edsel Ford, who had contributed generously to the center at Warm Springs. The NFIP was organized as a "membership corporation" (ie, a nonprofit organization); each member or donor was "enrolled to assist President Franklin D. Roosevelt and the Trustees in founding" the new foundation.[6] Its purpose regarding polio was fourfold: scientific research, epidemic first aid, education, and rehabilitation. Its original committee structure followed this plan. The NFIP mission statement that it would "lead, direct, and unify" the fight against polio nationwide ensured that it would supersede the local activities of the GWSF. FDR made it clear that he expected all fund-raising hitherto organized under the aegis of the Birthday Ball Committee to come under the control and supervision of the new foundation in 1938.[7] This, in part, was to deflect charges of partisanship that rankled the committee. Morgan and O'Connor, under pressure to orchestrate birthday ball fund-raising through the NFIP, found they could not legally begin solicitation under the name of the new foundation until the certificate was filed. This was done by January 3. O'Connor alluded to the principle of nonpartisanship just days after the NFIP started up, stating that "Everybody seems to realize already that the National Foundation for Infantile Paralysis is absolutely nonpolitical and that its only purpose is to combat one of the most frightening diseases that afflict mankind."[8] One year later, the President extolled the new foundation in a birthday broadcast as "a mature and efficient organization working industriously to perform its functions with but one objective—the banishment of infantile paralysis." The *New England Journal of Medicine* later echoed this perception in an editorial that "the work of the [NFIP] in three years has profoundly affected present-day knowledge of the disease by making a broad and continuous attack on many fronts."[9] Its scope of work, according to Morgan's planning advice in June 1938 that seemed almost fantastically hopeful at the outset included patient aid, research grants, iron lung services, monographs on polio diagnosis and care, grants to local hospitals and orthopedic centers, fellowships for medical research, and the establishment of district councils. The last idea would be supplanted by local county chapters, soon to come.[10]

[6] MDA, Medical Program Records. Series 10: Incorporation and By-Laws, 1938.
[7] FDRL, President's Personal File. Roosevelt to O'Connor. October 18, 1937.
[8] *New York Times*. January 17, 1938.
[9] FDRL, President's speech files, Box 34. *New England Journal of Medicine*, 22 (1942). p. 62.
[10] MDA, ibid.

Infantile Paralysis Fight, 1939. *Infantile Paralysis Fight* was the first program magazine for the president's birthday balls that shifted emphasis to the March of Dimes campaigns. The combined drive of 1939 was sponsored by the Greater New York Committee for the NFIP, chaired by Hugh S. Johnson, a prominent Brains Trust insider who headed the National Recovery Administration in 1933. This first and only number of *Infantile Paralysis Fight* transformed into *Courage Magazine* in subsequent years. The artist James Montgomery Flagg, best known for his Uncle Sam "I Want You" poster of World War I, contributed the cover art.

O'Connor launched into action at once to establish the new foundation on a firm footing by securing the top medical minds for NFIP medical committees. The first General Advisory Committee consisted of five members: Dr. Irvin Abell (President, American Medical Association);

Dr. Philip Lewin (Orthopedic Surgeon, Northwestern University Medical School); Dr. Thomas Parran (Surgeon General of the United States); Dr. Max Peet (Professor of Surgery, University of Michigan Medical School); and Dr. Thomas M. Rivers (Director, Hospital of the Rockefeller Institute). Paul de Kruif, bridging his participation from the President's Birthday Ball Commission for Infantile Paralysis Research, served briefly as Secretary of this committee until a falling out with O'Connor in 1941 over the denial of a grant to Dr. Thomas Spies provoked his abrupt departure. Before the break, O'Connor had asked de Kruif for general advice on selection of committee members and for particular assistance in recruiting Leo Mayer, Charles Lowman, and George Bennett.[11] He also asked de Kruif to approach JAMA Editor Morris Fishbein to serve on the Committee on Education. Fishbein became a stalwart advisor to O'Connor on many occasions. A. Graeme Mitchell, MD of the American Academy of Pediatrics Committee on Infantile Paralysis proved a forthright and sympathetic advisor, joining the committee a year later. In 1940, Fishbein joined the Committee on Medical Publications and quickly set standards for the committee's imprimatur on publications by NFIP grantees and committee members as well as term limits of committee members based on the AMA model.[12] Fishbein rose to prominence as O'Connor's closest consultant on committee operations, joining the General Advisory Committee which supervised the grant recommendations of the committees. He gave O'Connor his personal slant on the qualifications of physicians and the politics of medicine for nearly three decades, and O'Connor relied on him for his editorial talents and more. O'Connor was a bibliophile and a completist, and even his scholarly proclivities about polio were satisfied by Fishbein who, as editor of the massive *Bibliography of Infantile Paralysis, 1789–1944*, included O'Connor's tribute to the memory of FDR after his death. The dedication read: "Dedicated to Franklin Delano Roosevelt, 1882–1945/Who, by his triumph over the most dreaded of crippling diseases, which could not conquer him, gave inspiration and courage to thousands of children, men and women similarly afflicted."[13] While Fishbein provided O'Connor with a wide-ranging perspective on the medical profession at large, it was

---

[11] MDA, Public Relations Records. De Kruif, Paul.
[12] Ibid. Fishbein, Morris.
[13] Fishbein, Morris (ed.). *A Bibliography of Infantile Paralysis, 1789–1944*. Philadelphia: J. B. Lippincott Co., 1946.

Thomas Rivers of the Rockefeller Hospital above all that directed NFIP grants administration toward the most fruitful areas of research in virology. In sum, these committees and personnel formed the organizational framework that guided the NFIP in its first years to impel the scientific research on the causes of and solutions to polio.

NFIP Medical Advisors Meeting, 1942. Basil O'Connor poses with key advisors at the third annual medical meeting of the NFIP at the New York Academy of Medicine. [left to right] Commander Thomas M. Rivers, MD, Director, Hospital of the Rockefeller Institute for Medical Research and Chairman of the NFIP Committee on Virus Research; Basil O'Connor; Thomas Parran, MD, Surgeon General of the United States and member of the NFIP Committee on Epidemics and Public Health; and Morris Fishbein, MD, Editor of the *Journal of the American Medical Association* and Chairman of the NFIP Committee on Medical Publications.

With the NFIP established, it was left to Eddie Cantor (1892–1964) to give it a special, and unforgettable, cachet. Cantor was a multitalented vaudevillian, singer, actor, comedian, and radio personality whose rise to stardom began in 1917 in the New York City theatrical revue, the Ziegfeld Follies. Beloved by the American public and known as "banjo eyes" for his wide-eyed visage, Cantor embraced a host of humanitarian causes over the course of his long career in show business, including the March of Dimes and early support for the state of Israel. A high-energy show-man in both movies and radio also known as "the Apostle of Pep," Cantor was enormously popular for his song-and-dance routines and zany radio skits. On NBC's *Chase and Sanborn Hour* he debuted songs like "Santa Claus Is Coming to Town" (1934), an instant smash hit, and in musical comedies like *Whoopee!* (1930) he guaranteed his popularity in film for two decades to follow. On November 22, 1937, Cantor met with Woodbridge Strong Van Dyke II (assistant director in D.W. Griffith's silent film classic *Intolerance* in 1916) and Harry Mazlish of Warner Brothers in the office of John Considine, Jr. in the studios of Metro-Goldwyn-Mayer to discuss support for the new polio foundation. He recalled a successful radio appeal for relief funds after a catastrophic Mississippi River flood and proposed to help the new foundation through a similar radio pitch directed to the White House, pending the approval of the President. Cantor suggested, "We could call it the *March of Dimes*." This idea brought the approval of everyone in the meeting, but neither Cantor nor the others could have predicted how this simple phrase would catch fire to embody the fight against polio for years to come. Yet they immediately understood its appeal, based as it was on a pun on *The March of Time*, the contemporary Henry Luce newsreel known to every American who went to the movies. In actuality, *The March of Time* was first a weekly radio news series on the air from 1931 to 1945, a period spanning Roosevelt's presidency; it was followed avidly by millions of American listeners. The companion newsreel, also called *The March of Time*, ran from 1935 to 1951 to accompany feature films as an authoritative form of pictorial journalism that became America's regular immersion to the news of the day, which at the beginning of its run usually meant the brute realities of economic hardship and encroaching war. Cantor's *March of Dimes* idea was a brilliant stroke, and after the brainstorming meeting at MGM studios, the US comptroller J.F.T. O'Connor flattered Roosevelt that he had never seen "men who were more enthusiastic about anything as they were over the aid which they were anxious to render to disabled children." Cantor worked vigorously on the campaign by enlisting the support of MGM executive Nicholas M. Schenck, whose interest would soon prove critical, as well as popular entertainers of the day—Jack Benny, Bing Crosby, Rudy Vallee, Deanna Durbin, Lawrence Tibbett, Jascha Heifetz, Joe Penner, Kate Smith, and Edgar Bergen.

Eddie Cantor, 1946. An effervescent vaudeville star known as "Banjo Eyes" and "the Apostle of Pep," Eddie Cantor created the phrase "March of Dimes" and changed the direction of NFIP fund-raising almost overnight. Cantor's appeal for Americans to send dimes to the White House on his radio program in January 1938 captured the imaginations of millions, institutionalizing the philanthropy of ordinary people in the annual March of Dimes. In this photo, Mr. Cantor pinches a dime from Frances Rushmore, a John Robert Powers model perched atop a giant coin collection box for the 1946 campaign, as a portrait of FDR looms in the background.

The first *March of Dimes* radio appeal occurred during the week preceding the birthday ball events of January 30, 1938. On the air Cantor announced: "The March of Dimes will enable all persons, even the children, to show our President that they are with him in this battle against this disease. Nearly everyone can send in a dime, or several dimes. However, it takes only ten dimes to make a dollar and if a million people send only one dime, the total will be $100,000." This optimistic pitch collided head-on with the dismal news of a mere trickle of dimes to arrive at the White House in the days following the broadcast. But the trickle soon became a deluge. By January 29, over 80,000 letters (2,680,000 dimes or $268,000) flooded the White House mail room burying incoming correspondence in an avalanche of donations. On the eve of his birthday, President Roosevelt went on the air to express his thanks for the "countless thousands" of letters pouring in "literally by the truck load" to the White House. FDR admitted to his radio listeners that "I have figured that if the White House staff and I were to work on nothing else for two or three months to come we could not possibly thank the donors." This oft-repeated story of the first "March of Dimes" drive acquired the status of a myth of origins in later years. The reality of the situation was, of course, more complex than the airbrushed contours of the subsequent myth. In fact, there were two separate radio appeals, one by Eddie Cantor and the other by the Lone Ranger. According to Ira R. T. Smith, Chief of Mails at the White House, "the Lone Ranger emphasized the idea of putting a dime in an envelope and sending it to the President" and as a result the first wave of dimes was "largely from children." Smith confirms, however, that "the Government of the United States darned near stopped functioning because we couldn't clear away enough dimes to find the official White House mail." The situation Smith describes was unprecedented, even bordering on the ludicrous:

> Once we got 100 yards of dimes in scotch tape. … Other dimes were received in a wax design the size of a football. We had to boil it down. Thousands were baked into birthday cakes. Once we had to break up a concrete brick loaded with coins. A woman who wanted to contribute but had no money had her hair cut off and sent it in with the suggestion that we sell it. It brought eighty dimes. We received an aluminum cane made of hollow tubing into which were jammed 650 dimes. A gallon can containing $300 arrived from a bomber squadron. … One donor punched a hole in a dime (making it worthless) and tied it with twine to a huge cardboard tag labeled "The White House." I sent it around for Mr. Roosevelt to see. He kept the dime and returned the card with a new dime to go in the collection and a note saying, "I hope you have a good dime."[14]

The March of Dimes boosted incoming mail of 8,000 daily letters into the range of 50,000 to 150,000 daily. If this stupendous interruption of White House mail communications seemed like a one-off event at the time,

14 Smith, Ira R. T. *"Dear Mr. President…": The Story of Fifty Years in the White House Mail Room.* (New York: J. Messner, 1949). pp. 159–160.

it was not, for the "march" of dimes to the president continued year after year during FDR's tenure. Ira Smith stated, "After the first year, we made special preparations for the March of Dimes, taking on extra clerks in January and keeping them until we cleaned up the work in March."[15]

In jump-starting NFIP fund-raising as they did, Cantor and the Lone Ranger capitalized on the President's popularity and demonstrated the enormous practical value of an appeal for the inexpensive participation by everyone, young or old, rich or poor, simply by contributing a dime. Moreover, Cantor's knack of hitting on a catch-phrase that would be universally remembered was a stroke of genius, for the name *March of Dimes* as the NFIP's annual fund-raiser would soon become more widely recognized than the proper name of the foundation itself.[16] Eddie Cantor's radio appeal for dimes effectively circumvented O'Connor's original strategy of enrolling half a million founder members by soliciting 500,000 single dollar donations by subscription, much like the Polio Crusaders had done at Warm Springs a few years before. The "Fight Infantile Paralysis Campaign," styled as such by the new foundation, had opened with a radio broadcast in the Jade Room of the Waldorf Astoria on December 30, 1937, four scant days before the NFIP was officially organized. George M. Cohan (composer of two classics in the canon of American patriotic song, "Over There" and "You're a Grand Old Flag") was Master of Ceremonies, Dr. Max Peet offered a keynote address, and Basil O'Connor was the principal speaker. The *New York Times* published a lengthy article "National Attack Aims at Paralysis" by Paul de Kruif who explained how the NFIP would "utilize all the weapons of science" to battle polio. Women's committees organized teas, benefit theatrical performances sold out, and even Helen Keller was enlisted in the drive. As an "aerial parade" of aircraft buzzed over Manhattan, young women sporting red, white, and blue sashes and portrayed as debutante "Musketeers" strolled through Grand Central Station, restaurants, department stores, and hotels collecting donations in return for a tiny button bearing the slogan "*I'm Glad I'm Well.*" Frederick Snite, a well-known polio survivor popularly called "the boiler kid" and "the man in the iron lung," made a radio appeal from his iron lung respirator. This well-planned and auspicious beginning to the NFIP was then upstaged entirely by Eddie Cantor's "*March of Dimes*" radio pitch in the final week of January as

---

[15] Smith, Ira R. T. op. cit. See also Sills, David L. *The Volunteers: Means and Ends in a National Organization*. Glencoe: Free Press, 1957. p. 245.

[16] MDA, History of The National Foundation for Infantile Paralysis Records. Volume II, "Raising Funds to Fight Infantile Paralysis." Book 2, pp. 256–260.

newspapers delighted in photos of postmen straining under the weight of mail bags laden with dimes at the White House. Eddie Cantor continued through the war years as Chairman of the "March of Dimes of the Air," as the somber gravity of the NFIP gave rise to its public alter ego in the lively phenomenon of the *March of Dimes.*

Courage Magazine, **1942.** In the darkest year of the war *Courage Magazine* merged messages of patriotism, victory in war, and victory over polio. Though the mood was hopeful the situation was grim, and President Roosevelt's 60th (diamond jubilee) birthday might have gone unnoticed had it not been for the continuing January tradition of the birthday balls and March of Dimes. The youngsters on the cover, both professional models, were precursors to March of Dimes poster children, a program launched in 1946.

And a phenomenon it was. From 1939 to 1945, March of Dimes coin and birthday cards continued to be mailed to the White House as birthday greetings to the President. The cards were made available each year through March of Dimes county chapters, and there were often special distributions to retail chains and to the railroads. Grace Tully and other White House staffers intercepted countless off-season donations, rerouting them to NFIP headquarters in New York. Beginning in 1946, the year after Roosevelt's death, the cards were addressed strictly to local chapters. The popular visibility of the dime campaign was so great that one donor simply made a pencil rubbing of a dime so that its impression marked the only apparent "address" on the envelope. Not surprisingly, that dime got through to the President without a hitch, prompting a call to Ripley's "Believe It Or Not" to run the story.[17] The *March of Dimes* catalyzed the imagination, and fund-raising ideas proliferated. Mile O' Dimes installations appeared in 1939; these were wooden constructions enabling donors to place enough dimes end to end for one mile (about 92,160 dimes, or $9,216). The President's Birthday Committee issued instruction booklets to replicate this event anywhere. The March of Dimes made its way into Superman comics, John Wanamaker's, and Carnegie Hall. By 1943, with polio on the rise, the NFIP laid claim to over 40,000 volunteers as it began to involve grade school and high school youth in March of Dimes activities with Frank Sinatra as chair of the Youth Division. March of Dimes coin collectors and polio exhibits seemed ubiquitous, from storefronts and state fairs to the shop floor and Times Square, and FDR's portrait was usually a conspicuous point of reference in public displays. "*More in '44!*" was the campaign slogan in 1944 when NFIP media director Howard London organized a Jack Benny benefit tour, and the following year featured a full hour four-network show saluting FDR on his birthday, the last of its kind to be celebrated for its founder. O'Connor extended the March of Dimes drive till February that year due to snow storms that shut down participating movie theaters and continuing fuel shortages thanks to the war.[18] Above all, the early March of Dimes drives were a product of wartime America; the war was a mighty catalyst with an incalculable effect on the success of the organization.

At least one historian has claimed that World War II instigated a social revolution that was greater by far than the New Deal.[19] To understand this

---

[17] FDRL, President's Personal File, 4885.

[18] MDA, Fund Raising Records, Series 1, fund-raising appeals, 1945.

[19] Perrett, Geoffrey. *Days of Sadness, Years of Triumph*. New York: Penguin, 1973. p. 10.

claim, one must turn from the drama of military action and the tragic cataclysm of the war itself to examine the great transformation on the home front. First of all, the war definitively drew America out of the Great Depression as the frenzy of production overtook the economic doldrums of the previous decade. Preparations for war ensured jobs aplenty, and a vibrant culture of Swing Era jazz reached its peak as a mainstream and unifying musical force. Roosevelt himself sat at the helm as Commander-in-Chief, a revered but upbeat and time-tested leader. Many studies have documented Roosevelt's canny diplomacy in edging the nation closer to the confrontation with fascist aggression that he had recognized years before as an unmitigated evil and a danger to the very existence of civilization. The NFIP emerged in 1938 in an atmosphere of the crisis in Europe, with Adolf Hitler in the ascendant and the Anschluss, Munich, and Kristallnacht on the horizon. The subsequent popularity of the March of Dimes had much to do with FDR's own popularity in the evolving solidarity of the drift toward war and the developing culture of sacrifice by Americans at home with gasoline rationing and victory gardens emerging in response to the conflict.

Roosevelt went on the radio every year on his birthday to express his thanks to the millions who contributed to the March of Dimes and to reemphasize the importance of the polio campaign. Basil O'Connor and Eleanor Roosevelt did the same. In 1943, O'Connor and Mrs. Roosevelt greeted the nation from the White House just as FDR completed a war conference with Winston Churchill and Charles De Gaulle at Casablanca in North Africa. In his address O'Connor wove the themes of victory in war, Roosevelt's "four freedoms," and the March of Dimes, while Mrs. Roosevelt read her husband's succinct "thank you": "Please tell all of those who are helping so much in the great fight against infantile paralysis, that even though the visits I have been making in certain distant parts prevent my return to the Capital today, they are giving me once again a truly happy birthday. Tonight we are waging two wars, both in the service of humanity and both of them headed for victory."[20]

Despite the enormous burden of the war, Roosevelt and O'Connor continued their informal, Sunday-night meetings whenever their schedules permitted, O'Connor briefing the President as Roosevelt pored studiously over his stamp collection. They reviewed not only matters pertaining to the NFIP and GWSF but also much more. In one instance that conjures up the romantic intrigue of the film classic *Casablanca* (1942), the two collaborated on arranging safe passage for the son of a high-ranking Vichy official from

---

[20] FDRL, FDR Personal File, Speeches; 30 January 1943.

Casablanca to Warm Springs. Francois Darlan (1881–1942), Admiral of the Fleet of the French Navy, became Prime Minister under Marshal Petain in the formation of the Vichy government after France surrendered to the Nazis. Darlan's naval role in North Africa was formidable, and FDR attempted to win him over, dispatching General Robert Murphy in the autumn of 1942 to deal with Darlan as the Allies prepared to invade French North Africa. Darlan's son Alain, gathering intelligence for his father, suffered an acute attack of polio just at this juncture, and Roosevelt extended an invitation to remove him to Warm Springs. But before Admiral Darlan could respond to Roosevelt's offer of arranging his "retirement," he was assassinated. The young Darlan, immobilized by polio, was stranded with his family, but O'Connor arranged payment for their passage on the Isthmian Steamship line, and he arrived safely in Warm Springs for a year of treatment. One year later, FDR encouraged O'Connor to ask the Darlan family politely to leave Warm Springs since "their time is up" with his joking reassurance that an international incident over the eviction was then unlikely.[21] Yet the exigencies of fund-raising at home commanded O'Connor's attention far more than wartime geopolitics. In October 1943, he informed NFIP staff that the fourth war bond drive had been scheduled for January, coinciding precisely with the 1944 March of Dimes. He saw no reason to call off or alter the campaign; war bonds were just another complicating factor. Jokingly, he concocted the slogan, "Buy bonds and give us the change!"[22] But O'Connor was apt to complain to Roosevelt when political associates donated to other causes, as when New York State Racing Commissioner Herbert Bayard Swope distributed $628,294 of racing contributions to 14 charities but overlooked the NFIP entirely.[23]

In *The Life of a Virus* Angela Creager observed "there is a sense that the rhetoric of the NFIP *was* the rhetoric of war," and that "[t]he fight against polio thus embodied the transfer of both rhetoric and research organization from war mobilization to basic postwar science, a model made all the more powerful because the biomedical war against polio was 'won' by scientists in 1955."[24] The metaphors and rhetoric of warfare were indeed embedded in the language of the NFIP from its inception; with America's entry into the

[21] FDRL, President's Secretary File, 24 October 1943; April 1944. See also Peter Tompkins, *The Murder of Admiral Darlan*. New York: Simon and Schuster, 1965, and George E. Melton, *Darlan: Admiral and Statesman of France, 1881–1942*. Westport (CT): Praeger, 1998.

[22] MDA, Fund Raising Records, Series 1: Campaign Materials, 1944.

[23] FDRL, President's Secretary File. December 27, 1943.

[24] Creager, Angela N. H. *The Life of a Virus: Tobacco Mosaic Virus as an Experimental Model, 1930–1965*. Chicago: University of Chicago Press, 2002. pp. 179–180.

war, they became firmly entrenched. War metaphors were rife: *fight, conquest, battle, establish a beach-head, all-out attack,* polio as *the common enemy,* money as *fire-power,* research as a *weapon,* a girl rising up from her wheelchair pleading *Help me win my Victory,* the NFIP itself as *the first line of National Defense.* Moreover, certain metaphors were linked intrinsically to the specific circumstances of World War II: polio *attacks without warning* (Pearl Harbor; Nazi blitzkrieg), the insistence on *unconditional surrender* (FDR's end game strategy unveiled at the Casablanca conference), and the dark personification of poliomyelitis as *The Great Crippler* maiming and murdering the innocent resonated with an unconscious apprehension of Adolf Hitler as the incarnation of Evil and even with war itself as a cosmic manifestation of the Four Horsemen of the Apocalypse. At war's end, with the decisive triumph of Allies over Axis and Roosevelt in his grave, O'Connor proudly and deliberately remembered FDR as the indomitable world hero with words intended to hallow his memory for all time: "Polio Could Not Conquer Him." Even here, polio was a trope for everything else (eg, the Axis powers) that met its match in FDR. FDR himself contributed to the cross-pollination of martial rhetoric with metaphors about containing disease. On October 5, 1937, just days after he issued the statement creating the NFIP, he called for a "quarantine" against aggressor nations, stating, "When an epidemic of physical disease starts to spread, the community approves and joins in a quarantine of the patients in order to protect the health of the community against the spread of disease. ... War is a contagion, whether it be declared or undeclared."[25]

Employing the language of military mobilization to describe the program against polio was not irrelevant in wartime, for infectious disease and warfare shared some common territory. O'Connor pushed the idea that wartime conditions are conducive to the spread of epidemic disease and blamed the great 1916 polio epidemic on pestilence that arose during World War I. Further, he blamed the Japanese government's censorship of public discussion of polio a contributing factor to the spread of the disease in Japan, contrasting this with American freedom of expression and the righteous but the matter-of-fact stance of the NFIP.[26] Since it was the combatants, especially the wounded, who suffered most from such exposures in battle, O'Connor honored a real commitment to the armed forces via March of Dimes-funded science to military problems. In his speech "Scientific Research in the Emergency" he made it clearly explicit that the fight

[25] Philips, Cabell B. H. *From the Crash to the Blitz, 1929–1939.* New York: New York Times. p. 539.
[26] MDA, Basil O'Connor Papers. Series 1: Typescripts. Radio address, January 14, 1943.

against polio would lead to the conquest of other diseases, and indeed the Foundation had placed the investment in basic science as a cornerstone of its research grants program. O'Connor also saw the rising incidence of polio linked to a decrease of productivity in factories geared to maximum production. According to the NFIP, "research and more research" was the key weapon that would bring the disease to its end.[27]

The American public learned about, and participated in, the March of Dimes in the social circumstances of global war. In 1941, before America's entry into the war, the NFIP created a birthday coin card addressed to *President Franklin D. Roosevelt, White House, Washington, DC* that enlisted each donor in a neighborly, patriotic effort: "I have joined our National Defense against Infantile Paralysis to help the youngsters around our own corner." Whether this dime card can be interpreted as O'Connor's shrewd attempt to help shape public opinion to abet FDR's maneuvers against isolationism before Pearl Harbor is uncertain, but the message was a civics lesson and a sign of the times. In a radio address that year, FDR helped the American public to understand the NFIP in this way:

> I have always tried to remember that the particular problem of infantile paralysis does call for a truly national fight. We had it in every state of the union. We are at last organizing adequately to fight it. We have had to face the necessity of uniting medical scientists and doctors and nurses and public health officers and the general public into a unique offensive and the battle year by year is gaining success.[28]

After Pearl Harbor, March of Dimes chapters had no hesitation in uniting the causes of polio and the war emergency in ultra-patriotic appeals on behalf of the Commander-in-Chief. In one case, the Alabama Committee for the Celebration of the President's Birthday crafted a letter with expressions of devotion to "this true man, our President Franklin D. Roosevelt" by quoting General Douglas MacArthur before his escape from the Philippines under attack by the Japanese:

> Today, January 30, 1942, the Anniversary of your birth, smoke-begrimed men, covered with marks of battle, rise from the fox holes of Bataan and the batteries of Corregidor to pray reverently that God may bless immeasurably the President of the United States. ... Ask yourself WHAT DID YOU DO on That Day when those men in foxholes fighting for their lives stopped long enough to send a greeting to our President. If you did not send a remittance on that day, do it today![29]

---

[27] MDA, Basil O'Connor Papers. Series 3: Printed Literature. Speeches and addresses, May 20, 1942. Fund Raising Records, Series 1: Campaign Materials. White House luncheon, Mary Pickford broadcast, 1944.

[28] MDA, Radio Spot Announcements Records. FDR radio speech, 1941.

[29] MDA, Fund Raising Records. Series 1: Campaign Materials. Alabama Committee, 1942.

The chapter letter ended with the cocky directive, "Please throw this letter away, if you are not 100% behind our great Commander-in-Chief." January 30, 1942 marked Roosevelt's 60th birthday, and the Diamond Jubilee birthday balls celebrated with muted pageantry across the nation to benefit the March of Dimes repeated the cry, "Remember Pearl Harbor!" In a speech on March 3, O'Connor reflected that the President's Diamond Jubilee Birthday "gave proof … that the 'V' to vanquish disease is synonomous [sic] with the 'V' for victory," and he peremptorily dismissed nay-sayers as "Fifth Columnists."[30]

The losses at Pearl Harbor became a benchmark by which all casualties were measured. Polio claimed 12,000 victims in 1943, "more than three times the number of casualties at Pearl Harbor," in the words of a press release from the NFIP. The media messages resounded with the realities of the war: "We are prepared to fight [polio] with the planned strategy of a military campaign—not only because the enemy is a merciless and insidious one, but because the danger of epidemic in wartime makes this fight an actual military necessity." In his radio broadcast for the 1944 March of Dimes, O'Connor reiterated the Rooseveltian war goal, insisting that the "unconditional surrender" of polio was the ultimate end: "I'd like to tell you a few things about infantile paralysis so you'll realize, as I do, that the fight against it is *every* one's fight. It's not a very pleasant subject. And neither is war. But they've both got to be faced fairly and squarely—and then licked into *un*conditional surrender."[31] In turn, Roosevelt's complementary address at the time appreciated the "frontline fighters" of the March of Dimes: "The tireless men and women working night and day over test tubes and microscopes, searching for drugs and serums for methods that will prevent and cure—these are workers on the production line in this war against disease. The gallant chapter workers … [and] the volunteers who go into epidemic areas to help the physicians—these are the frontline fighters."[32] In his last public message about the March of Dimes in 1945, Roosevelt brought together all the familiar themes with his vision of American democracy, triumphant and polio-free:

> *The success of the 1945 March of Dimes in the campaign against infantile paralysis does not come as a surprise to me. We are a nation of free people, and free people know how to go over the top—whether it's a Nazi wall, a Japanese island fortress,*

[30] MDA, Fund Raising Records, Series 1: Campaign Materials. Correspondence, 1942. Basil O'Connor Papers, Series 1: Typescripts. "The Challenge of Tomorrow," 1942.

[31] Ibid. Press releases, 1944. Radio broadcasts, January 23, 1944.

[32] FDRL, President's Secretary's File. Basil O'Connor, 1942–1944.

a production goal, a bond drive, or a stream of silver dimes. The reason for these achievements is no military secret. It is the determination of the many to work as one for the common good. It is such unity which is the essence of our democracy.[33]

By the time of his death, not only the American people but millions around the world had heard FDR's voice on the air introduced by radio announcers whose steely exhortations buzzed with electricity: "*Report to the Nation! Infantile paralysis is on the march! Help fight infantile paralysis now— before it attacks your children! Money is the ammunition we need! Send your dimes and dollars to President Roosevelt at the White House! It will hurt until you give. Join the March of Dimes—today!*" But when global hostilities ceased the polio war was far from over: "*Humanity declares war on infantile paralysis!!! And the spirit of Franklin Delano Roosevelt still leads the fight to save our children from this crippling disease.*"[34]

Through the war years, the March of Dimes drives encompassed two weeks of intensive fund-raising culminating in the birthday celebrations on January 30. Despite FDR's tacit approval for the use of his birthday, it became a matter of tradition for O'Connor and Morgan to seek his permission each year. This formality took on added significance with direct involvement of United States in the war, for the campaigns should not appear to interfere in any way with war mobilization. O'Connor worked to dispel the rumors that fund-raising would be called off in any year due to the exigencies of war; he was always apt to dismiss rumor-mongering as a business intrusion. In 1942, he drafted a two-page letter to be discussed with FDR at the White House with his rationale for the 1943 March of Dimes; Roosevelt's formal reply set forth his own justification of maintaining the drive, even in the circumstances of the war. The importance of this and similar letters in the war years, undoubtedly a formalized exchange planned beforehand, cannot be overestimated (see Appendix I). Roosevelt's authorization of his birthday for the March of Dimes contextualized the moral integrity of the NFIP in terms of American war goals while simultaneously justifying the Foundation's humanitarian mission on its own terms. FDR's rationale thus served to prevent criticism that fund-raising for polio was frivolous, secondary, or extravagant if compared to preparations to meet the global war emergency. So important to FDR was this justification that he requested additional copies of the letter from O'Connor.[35] The exchange

---

[33] FDRL, FDR Speech Files. Speech, January 30, 1945, read by Eleanor Roosevelt.

[34] MDA, Media and Publications Records. Keystone Broadcasting System announcement, 1944. Foreign language radio spot announcements, 1945–1946.

[35] FDRL, President's Secretary File, Basil O'Connor, 1942–1944.

was reproduced in full in the *National Foundation News* and 1943 campaign handbook, and O'Connor arranged for both letters to be printed in a pamphlet with the title, "We must help them win their Victory over disease today," quoting from Roosevelt's letter. Again, "Doc Roosevelt" of Warm Springs and "Doctor Win-the-War" of the White House map room converged in the unified persona of the Commander-in-Chief who had personally "conquered" polio and was bent on its conquest for all Americans. O'Connor enshrined one FDR quotation from the letter in NFIP literature thereafter: "Nothing is closer to my heart than the health of our boys and girls, and young men and young women. To me it is one of the front lines of our National Defense." This statement articulates one of the core values of the NFIP during wartime, perfectly complementing its mission statement to "lead, direct, and unify" the fight against polio. Roosevelt's expression combines the warmth of parental concern with the patriotic impulse to preserve American democracy, a shrewd but touching lesson that O'Connor would have repeated time and again. The message was often seen as a caption on posters with FDR's portrait.

O'Connor also repeated the gesture of seeking permission for the FDR birthday galas one year later with similar rhetorical flourishes. He alluded to the dramatic upswing in polio cases in 1943, noting that "unavoidable wartime crowding and mass movements of people" increased the risk of contagion. O'Connor grew fond of claiming that the NFIP was "owned and maintained by the American people," a connection that he never failed to emphasize when addressing any audience in years to come. Without exception, in speech and in print, he religiously referred to the NFIP as "*your* National Foundation," an essential part of participatory democracy. The war on polio was presented as democracy in action, uniting the forces of an informed public with mission-oriented science. To FDR, he concluded again in 1943: "In a fight against infantile paralysis, your birthday is much more than a date. It is a symbol. I hope we may use that symbol."[36] O'Connor well knew the potency of this symbolism, and FDR's response was predictable: "The threat of this baffling, unconquered disease still hangs ominously over our vital war effort. There can be no armistice with the Crippler. Surrender of disease on the home front must also be unconditional." Further, FDR situated the fight in the widest public health context possible; he wrote "While our men and women on the battlefields of war are fighting

---

[36] FDRL, President's Secretary File, Basil O'Connor, November 4, 1943.

for the Four Freedoms, we at home must continue our fight against another enemy—Disease." In the 1946 March of Dimes, O'Connor would finally link the NFIP to the Four Freedoms explicitly, announcing a Fifth Freedom—Freedom from Disease.[37] This concept, conjoined with the famous Rooseveltian principles of the Four Freedoms, formed the philosophical underpinning of the entire NFIP program, not only in its support of research in basic science above and beyond the polio mission, but also in its future transformation to a leader in birth defects prevention.

US Army Neurotropic Virus Commission, 1943. John R. Paul, MD, Yale University; Major C. E. Van Rooyen, Royal Army Medical Corps; Major Albert Sabin, MD, US Army Medical Corps; and their chauffeur at a field laboratory hospital near Cairo, Egypt. As members of the US Army Neurotropic Virus Commission under a grant from the NFIP, Drs. Paul and Sabin investigated the causes for polio outbreaks in American troops in North Africa when the disease was almost entirely absent among native Egyptians. This was an example of the so-called "shoe leather epidemiology" conducted by the Yale Poliomyelitis Study Unit founded by Dr. Paul that pioneered clinical epidemiological studies of polio in small communities.

Anticipating the nation's wartime needs, O'Connor himself had already pondered the issue of an intrinsic link between polio and military matters

---

[37] FDRL, President's Secretary File. November 8, 1943. MDA, Public Relations Records.

as early as 1940, asking Morris Fishbein, "How can the National Foundation tie in with the Army and Navy in doing a job in connection with
defense on infantile paralysis? Isn't that a very simple question?"[38] His query
to Fishbein ultimately led to specific projects funded by the NFIP during
the course of the war, but the patriotic rhetoric that ensued after Pearl Harbor generalized the NFIP–military connection. Beyond the diplomatic gesture of aiding the Argentine polio epidemic in 1943, the NFIP soon found
specific ways to support the military that advanced real knowledge of poliovirus. In 1943, the NFIP approved a grant of $15,000 to the US Army
Commission on Neurotropic Virus Diseases for an expedition to North
Africa by Dr. John Paul and Dr. Albert Sabin to investigate not only poliomyelitis but also encephalitis, infectious hepatitis, and sandfly fever occurring in Egypt and the Middle East.[39] Besides Egypt, the team traveled to
Palestine, Iraq, Iran, India, Sudan, Eritrea, Libya, Algeria, Tunisia, and Sicily
in the wake of Allied military success in North Africa. The commission
concluded that although the disease was not highly prevalent among American troops, it posed a problem for military personnel in the Mediterranean
and the Middle East and that its appearance in a sporadic rather than epidemic form in Egypt had diverted medical attention from its virulence and
spread.

From this point forward, the NFIP enjoyed many productive relationships with the US military, notably a standing agreement with the US Air
Force known as the Military Air Transport Service (MATS) for the air
transport of polio patients in iron lungs to hospitals throughout the nation.
NFIP relations with US Armed Forces benefited through the course of the
war by O'Connor's standing with the President, and the NFIP fund-raising
department created an "Armed Forces Division" that reaped the benefits of
this connection for years to come. In 1943 alone, military personnel donated
over $100,000 to the March of Dimes as polio cases peaked in the second
largest epidemic since 1916. In the interest of supporting the March of
Dimes and preventing polio epidemics, endorsements of top military brass
were never a problem. In seeking military endorsements, O'Connor never
hesitated to go straight to the top. For the 1944 March of Dimes, Gen.
Dwight D. Eisenhower, US Army Supreme Commander of the Allied
Expeditionary Force in Europe, sent this telegram message in response to
O'Connor's personal appeal: "Our soldiers in the field make no distinction

[38] MDA, Public Relations Records. Fishbein, Morris. August 14, 1940.
[39] MDA, Committee on Research—Basic Science Grants Records. NYSA, Gen. James Magee to
O'Connor. March 23, 1943.

in their minds and hearts among the enemies that imperil the safety[,] integrity[,] and well being of our nation. We recognize infantile paralysis as one of these enemies and are anxious to do our part in eliminating it. I am certain that I voice the sentiments of this entire command when I say that the anti-infantile paralysis campaign is in a very definite sense entitled to be classed as another defender of this country."[40] General Eisenhower promised that his personal contribution to the March of Dimes would follow the telegram, and it did. Such charitable commitments from the top brass of the Army and Navy made for good public relations, and they were matched by generous donations from armed forces personnel dispersed across the globe in the circumstances of the war. By war's end O'Connor's prominence as the civilian "general" marshaling the resources of charitable giving would be further expanded by his role at the head of the Red Cross coordinating relief efforts. The "Armed Forces March of Dimes" became a mainstay of NFIP fund-raising through the 1950s, and even until the late 1960s the Foundation kept a retired officer on staff as liaison to the military. Aerial photo stunts depicting sailors in formation on the deck of the US aircraft carrier *Franklin D. Roosevelt* spelling out *MARCH OF DIMES* continued to recall the connection between FDR, America's armed forces, and the March of Dimes, but simpler, everyday situations of GIs dropping a dime in a collection can for crippled children were common enough. In their very last photo opportunity to appreciate the inventiveness of military campaigns for the March of Dimes, Roosevelt and O'Connor met with officers of the Persian Gulf Command at the White House on March 21, 1945 to accept an unusual donation for the March of Dimes. The command was a US Army service command that ensured the transport of lend–lease materiel along a supply line through Iran to the Soviet Union. To support the 1945 March of Dimes, one supply run was designated "the March of Rials Special" (the *rial* is the currency of Iran) that made a 570-mile journey from Teheran to the Persian Gulf collecting $19,438 from servicemen at way stations along the route.[41] The officers presented a check for this amount to FDR at a small ceremony at the White House, just 2 weeks before his death in Warm Springs.

It remains difficult to fathom the enormity of the catastrophe that World War II brought to humanity. More than 55 million people died in combat or as a direct result of hostilities. The *March of Dimes* was born in the crucible of that great conflict, and though it may have been inconspicuous in

[40] NYSA, Basil O'Connor Records. Dwight D. Eisenhower to Basil O'Connor. January 24, 1944.
[41] *National Foundation News*, April 1945. 4(6): p. 3.

the overall course of the war itself, it was a regular mainstay of home front activities, congruent in every way with the high-minded spirit of patriotism, sacrifice, and commitment to the public health of the entire citizenry that the war demanded. The March of Dimes fluidly channeled patriotic feelings about the Commander-in-Chief through a philosophy of volunteerism that captured Basil O'Connor's aspirations of creating a national civic organization that would put a stop to polio. It continued to build on that spirit of volunteerism after the war, and its momentum carried it into the very heart of American popular culture. In August 1945 the war ended with the Japanese surrender. Just at that moment, the city of Rockford, Illinois in Winnebago County was hit with a polio epidemic that had by its end paralyzed 382 individuals, most of them children under 15, and killed 36. There were 27 families with multiple cases of polio. The Winnebago County Chapter of the NFIP set up an emergency headquarters in a downtown hotel and issued this stern warning to Rockford: "The Infantile Paralysis Committee strongly recommends that parents keep their children under 14 from congregating in parks, playgrounds, business districts and other public places. It is important that children be kept from making new contacts and from engaging in all fatiguing activities. Insofar as possible, children should be kept within the areas of their own yards." The committee also put out an appeal for nurses to help manage the overload of cases. The US Army, the NFIP, and the Red Cross all sent nurses to Rockford, but many unaffiliated nurses also heeded the appeal. One said, "While preparing my evening dinner at home in New Brighton, Minnesota, I heard an appeal for nurses in Rockford, Illinois over radio station KSTP during their news broadcast August 29. Two days later I was working in Rockford."[42] By late August the epidemic seemed to abate, but then nine new cases were reported on August 25, in part attributable to the "spontaneous congregating of thousands of persons on downtown streets" to celebrate the end of the war 10 days earlier.[43] As then estimated, 10 days was considered the typical incubation period of poliovirus prior to developing symptoms, and the revelry of Rockford's citizens, who shared with the world their jubilant gratitude that hostilities had ended, had seemed to enhance the spread of the polio contagion in their city. World War II had come to an end, but polio had not.

---

[42] MDA, Publications Collection. Winnebago County Chapter National Foundation for Infantile Paralysis, Faust Hotel. Rockford, Illinois. 1945.

[43] *Rockford Republic*, "Blame V-J for Polio Flareup." August 25, 1945.

Polio: Unconditional Surrender, 1947. The early years of the March of Dimes coincided with tensions in Europe and Asia that led to the catastrophe of global war. Americans first experienced the annual March of Dimes campaigns in the context of World War II, and the metaphors of organized warfare and the rigors of battle were translated into energetic slogans and cartoon propaganda about defeating polio. This dark cartoon released after the war in 1947 personifies the demonic evil of polio conflated with the Axis powers in the image of the Grim Reaper as the March of Dimes issues an ultimatum of "unconditional surrender," a Rooseveltian principle that was the backbone of the Allies' approach to the menace of Nazi Germany.

CHAPTER 6

Hollywood and the Publicity Machine

In 1938, as the March of Dimes burst upon the scene with great fanfare, it capitalized on every popular entertainment of the Swing Era from Hollywood to Yankee Stadium. The early March of Dimes campaigns, held every January following the tradition of observing Franklin D. Roosevelt's (FDR) birthday with music and dancing to fight polio, were replete with endorsements from celebrities in the worlds of entertainment, the arts, sports, and politics. The fight against polio ran parallel to America's mounting opposition to totalitarian dictatorship—everything that was wholesome and cherished in American culture, especially the protection of children from disease, corruption, and violence, could be captured in the symbolism of the March of Dimes. The campaign was emblematic of the partly acknowledged vulnerability of an extremely popular president, who was seen to have recovered from his polio disability to regain the vitality of a normal adult. In the rising confrontation between the Allied and Axis powers, the March of Dimes put forward every public virtue in slogans that might differentiate America from the brutality of dictatorship and war—neighborliness ("help the child around *your* corner"), charity ("your dime will help a child walk again"), volunteerism ("*Join* the March of Dimes"), cheerfulness ("Make Giving a Game"), and public health ("I'm Glad I'm Well"). These virtues were amplified time and again not only in slogans intended to educate about polio, but by the celebrities who jumped aboard the bandwagon because they admired FDR. Their fame and visibility helped advance a cause whose patriotism and affirmation of life and health were unquestionable. Even when war came to America in 1941, the polio fight was not put out of service but enlisted alongside Roosevelt's grand campaign to win not only the war but also to restore peace and health to the world.

In 1939, the year after Eddie Cantor's famous "*March of Dimes*" radio pitch, as the world shuddered at the Nazi onslaught, the bifurcated endeavors of birthday ball events and March of Dimes activities merged (as Roosevelt had wanted), only to begin to diversify by countless ideas at the local level to support both aspects of the unifying drive. New ideas had been

Friends and Partners
ISBN 978-0-12-803597-9
http://dx.doi.org/10.1016/B978-0-12-803597-9.00006-0

encouraged during the birthday balls during the Depression years, and were given further encouragement now under the direction of the National Foundation for Infantile Paralysis (NFIP). Keith Morgan and Basil O'Connor deployed their Hollywood and business connections to ramp up the excitement. The "*Infantile Paralysis Fight*," as the March of Dimes campaign was positioned in 1939, saw support from all three major broadcasting networks—Columbia, Mutual, and National—in 30 special radio broadcasts for the 1939 March of Dimes. These broadcasts featured appeals from Bob Hope, Fanny Brice, Morton Downey, Mickey Rooney, Norma Shearer, and Betty Jaynes, all well-known personalities of the time. In New York City, the Greater New York Committee of the NFIP organized the Waldorf–Astoria birthday ball where movie stars and bandleaders stepped forward as "The Entertainment Brigade." Dancers Mayris Chaney and Eddie Fox created a new ballroom dance step, the "Eleanor Glide," in honor of Mrs. Roosevelt, and the First Lady was charmed. The committee chair for the campaign was General Hugh S. Johnson, a cantankerous and controversial figure who had headed FDR's National Recovery Administration (NRA) in 1933. From mainstream sports Lou Gehrig, Babe Ruth, and Jack Dempsey added the extra clout of their support for the NFIP, and New York City Mayor Fiorello LaGuardia gave his blessing over all. The Russian pianist Simon Barere performed with Lily Pons at a benefit concert on January 25 at Carnegie Hall. The synergy of these events increased the excitement, and the dimes poured in.[1]

In the years during the war, as the sophistication of the Foundation's approach to the annual campaign evidenced both the organizational development of the New York headquarters and an expansive network of local chapters nationwide, the March of Dimes became an inspirational ritual that colored the privations of the war with a strong tinge of hope and the healthy distraction of entertainment. True, the local birthday balls for the president continued across the nation, but the campaign assumed a powerful new form in the matrix of mass media; that is, radio and movies (television would not come to dominate the media until after the war). By 1941, Helen Hayes (1900–93) had lent her name to the cause of polio prevention. Widely known as the "First Lady of the American Theater," Hayes was a media phenomenon yet exceedingly down-to-earth, an actress whose crossover success from film to Broadway to radio (and later to television) endeared her to the sophisticated and to the masses. She supported the March of

---

[1] *Infantile Paralysis Fight*. New York: The National Foundation for Infantile Paralysis, 1939.

Dimes and other charitable causes with regularity, but in 1949 when her daughter Mary MacArthur died at age 19 after a bout with polio, she redoubled her efforts to aid the NFIP. Hayes was not a figurehead celebrity by any means: her lasting friendship with Basil O'Connor and Elaine Whitelaw (NFIP Director of Women's Activities) led to a conspicuously high-profile role as the first national Mothers March chair from 1951 to 1961 after "Mothers March against Polio" (a door-to-door solicitation campaign) entered the picture in 1950, and she established the March of Dimes Mary MacArthur Fund in memory of her daughter.[2]

The war years drew forth other famous personalities as well: Tyrone Power, Humphrey Bogart, Maureen O'Sullivan, and Dorothy Lamour from Hollywood; Jack Benny, Fred Allen, Lowell Thomas, William Shirer, Eric Severeid, and Mel Allen from journalism and broadcasting; Benny Goodman, Kay Kyser, Connee Boswell, Harry James, and Glenn Miller who were Swing Era jazz leaders universally known; and Marjorie Lawrence, Lauritz Melchior, and the Maestro Arturo Toscanini from the worlds of opera and orchestral music. Among these, the careers of Connee Boswell and Marjorie Lawrence were directly affected by polio. A soprano famed for her interpretations of Wagnerian opera, Lawrence's career was cut short by polio in 1941; her story was later told in the film *Interrupted Melody* (1955). Connee Boswell's disability was probably caused by an accident and not polio, but she performed publically in a wheelchair and within the ambit of pop culture remained closely associated with the March of Dimes from its inception. During this period of growth and ferment, several representative roles were characterized as honorific titles in the annual March of Dimes: Glenn Miller was the Chairman of the Dance Orchestra Leaders Division, Eddie Cantor was the Chairman of the March of Dimes of the Air Committee, Grantland Rice was the Chairman of the Sports Council, and Frank Sinatra became the Chairman of the Youth Division.[3]

Prominent graphic artists and illustrators, though perhaps not as widely known as big name Hollywood stars, also made vital contributions. McClelland Barclay created the poster "Fighting in the Dark," which recapitulated bandleader Artie Shaw's hit tune "Dancing in the Dark." Barclay was best known for his *Saturday Evening Post* magazine covers and for the General Motors ad campaign "Body by Fischer." Lucille Patterson Marsh, illustrator of children for *Ladies' Home Journal*, General Electric, and Ivory

---

[2] MDA, Women's Activity Records. Series 3: National Chairpersons. Helen Hayes, 1949–1969.

[3] *Courage*. New York: The National Foundation for Infantile Paralysis. Volume 1, 1941; Volume 2, 1942; Volume 3, 1943; Volume 4, 1944.

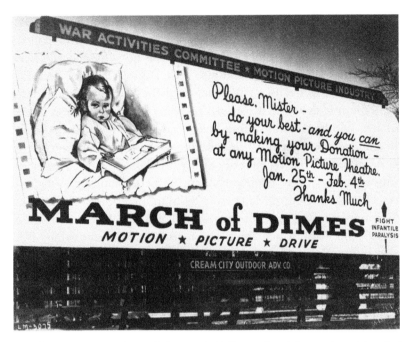

March of Dimes Motion Picture Drive, 1943. Hollywood and the motion picture industry conspicuously supported the March of Dimes during the war. Stars such as Mary Pickford, Douglas Fairbanks, Jr., and Greer Garson made appearances for the March of Dimes, while the film studio executive Nicholas Schenck curried favor with FDR by supporting dime collection programs in movie theaters nationwide. The War Activities Committee of the motion picture industry sanctioned and mobilized this fund-raising. The artist Lucille Patterson Marsh contributed the artwork used in this billboard advertisement, a precursor to the poster children of later years.

Soap, drew a piteous image of a little girl in bed recovering from polio that appeared in March of Dimes literature and roadside billboards, an inspiration for scores of March of Dimes posters that flooded the popular landscape with images of children in recovery from polio in the 1950s. James Montgomery Flagg, famous for the Uncle Sam "I Want You" poster of World War I, contributed his artwork to the covers of the official magazine programs for the grandiose birthday balls at the Waldorf Astoria, including *Infantile Paralysis Fight*, the NFIP birthday ball magazine of 1939. Flagg's artwork for the 1938 cover, "The Dawn of the New Foundation," featured his granddaughter and her playmates under an allegorical figure of liberty. This 1938 magazine was possibly the finest combined expression of Roosevelt's and O'Connor's project for popular consumption at the very birth of the NFIP. Together they announced a complete change of tactics in the unification of all antipolio activities under the new foundation. FDR's statement on the creation of the NFIP was reprinted as well as his letter of thanks to O'Connor for his leadership. O'Connor's brief introduction, "For Greener Fields," established FDR as responsible for the "whole movement" and set forth their goal of $5 million in $1 contributions from 5 million donors, or "members," before the end of 1938.[4] Photos of the entire Board of Trustees were displayed, and endorsements from top celebrities were overflowing with support of FDR's new idea. These included the comedy duo George Burns and Gracie Allen, Eddie Cantor, Kate Smith, James Cagney, Bette Davis, and even the Marx Brothers, who wrote:

> It's every man's privilege to differ with the President of the United States on political matters. But no man can differ with FDR's unselfish battle to stamp out infantile paralysis. His splendid leadership here deserves the loyal support of every good American. When little boys and girls need help, politics should be forgotten. The new National Foundation for Infantile Paralysis can become a power for good in this country comparable only to the American Red Cross.[5]

Why such an overwhelming outpouring of support by the greatest artists and entertainers of the era? Four intertwined strands of influence potentiated the collaboration of celebrities with the NFIP: the popularity of FDR as a "champion of freedom," the global reach of broadcast media, the patriotic momentum of the nation in the developing context of war, and the charitable impulses of the millions who agreed that poliomyelitis was a monster that deserved extinction. It called for the genius of Basil O'Connor to find a convincing marketing vehicle to pull all the pieces together, and he did so not through any messianic posturing or even from the charisma he may have borrowed from FDR, but as an exceptionally capable manager convinced that the nation would find a solution to polio provided the critical medical and social resources were effectively coordinated. O'Connor operated from the general model of the New Deal: experimentation and action. And, like Roosevelt, he succeeded by effective delegation of responsibility to experts in every field.

The person most responsible for the Foundation's continuing success in utilizing celebrity talent was Howard J. London. O'Connor discovered London as a talent buyer for a small radio agency, and he quickly rose to the role of director of the NFIP Radio, Television, and Motion Picture Department in the 1940s. Through his connections with celebrity agents in New York, Chicago, and Los Angeles, he ran the March of Dimes campaigns

---

[4] DART, Basil O'Connor Papers. *President's Birthday Ball Magazine.* January 1938, 1(3):2.
[5] Ibid, p. 44.

as show business, explicitly intended to entertain. London insisted on dealing only with the top names in entertainment and arranged benefits, tours, radio spots, and movie trailers with familiar figures such as Jack Benny, Doris Day, and Bob Hope. His department later produced educational films about polio and the NFIP, such as *Your Fight against Infantile Paralysis* (1945) and *In Daily Battle* (1947). London was not always a congenial colleague; his ideas and methods clashed with those of Public Relations Director Dorothy Ducas, and their intramural wrangling sometimes led to petty conflict over who commanded the turf of publicity at the NFIP. In the end, O'Connor trusted them both, though it was the insight of Ducas that he relied upon most when publicity about the Salk vaccine was at stake in the 1950s. Howard London's creativity brought forth a wide range of novelty "stunts" (as they were then called): the Miss Hush radio contest (choreographer Martha Graham played "Miss Hush"), floats in the Tournament of Roses Parade, and the RCA Starliner Tour, a cavalcade of musicians traveling by rail; all of these were March of Dimes publicity ideas.

The creation of the NFIP on January 3, 1938 came midway between two of the most notable musical events of the time, just a few weeks apart. The first was the inaugural concert of the NBC Symphony Orchestra conducted by the world-renowned Arturo Toscanini on Christmas Night, December 25, 1937; and the second was the Benny Goodman Orchestra concert at Carnegie Hall on January 16, 1938. These were pivotal musical events the importance of which cannot be overestimated. The NBC Orchestra was created specifically for Maestro Toscanini, and it went on to an illustrious 18-year career under his baton as the "Symphony of the Air." The radio program reached millions of listeners in the United States and Canada. The Benny Goodman engagement at Carnegie Hall has been hailed as the first significant breakthrough of a jazz ensemble at this mid-Manhattan palace of high culture and classical music, and it solidified Goodman's reputation as the "king of swing." Swing music energized the nation during the 12 years of Roosevelt's presidency as the most popular time in the history of jazz: this art form was indeed the cultural expression of the New Deal. Hundreds of big bands toured the nation to bring music and dancing to the masses, but New York City was the very center of this explosion. The March of Dimes suddenly appeared in the midst of this musical ferment, when jazz was *the* pop music of America. The annual president's birthday balls from 1934 to 1945 were predicated on music and dancing, regular occasions when the geniality of FDR, the charitable impulse to stop polio, and the effervescence of swing coalesced in the activities of a single organization (first the Georgia

Warm Springs Foundation, GWSF, from 1934, then the NFIP in 1938). The March of Dimes as a popular mass phenomenon enlisted the energy of jazz just as it did the glamour of Hollywood to empower charitable giving with entertainment. Swing was uplifting with syncopated energy, and it is no surprise that specific popular songs lent a symbolic identity to the beginning of Roosevelt's presidency ("Happy Days Are Here Again") and to Eddie Cantor's famous "March of Dimes" pitch in 1938 with the celebrated hit by Irving Berlin, "Alexander's Ragtime Band." FDR and the New Deal were memorialized in dozens of blues and folk songs of the period. With Ira Gershwin's swing song "I Can't Get Started With You," Billie Holiday crooned, "*I've been consulted by Franklin D./Robert Taylor has had me to tea/But now I'm broken-hearted/I can't get started with you.*" There were also popular versions by Frank Sinatra and Bunnie Berigan. FDR's 1936 visit to the Caribbean endeared him to many as a prophet of peace as evidenced in the song "FDR in Trinidad," and two decades later a calypso band emerged that called itself "The March of Dimes Quartet." Roosevelt deeply appreciated the power of musical performance; on one special occasion he reached out with personal gratitude to Arturo Toscanini for conducting a benefit concert for the NFIP. In a letter of April 6, 1943 FDR's *thank you* on behalf of "crippled children" moves fluidly into a tacit but significant acknowledgment of Toscanini as an ardent antifascist and opponent of Mussolini just as Allied forces prepared to invade Sicily in a new phase of the war.

> My Dear Maestro—
> It is with great personal pleasure that I convey my deep appreciation to you and the members of the NBC Symphony Orchestra for your brilliant concert dedicated so generously to the work of the National Foundation for Infantile Paralysis.
> The crippled children for whose sake you gave us an evening of inspiring music shall never forget you. They have derived new hope and courage in their fight for health and happiness.
> The magnificent contributions you have made to the world of music have always been highlighted by your humanitarian and unyielding devotion to the cause of liberty. Like all true artists you have recognized throughout your life that art can only flourish where men are free. Once again, your baton has spoken with unmatched eloquence on behalf of the afflicted and oppressed.
> Even while the United Nations are engaged in a momentous struggle to preserve the culture and heritage of free peoples everywhere, it is heartening to know that our fight against infantile paralysis continues unabated, thanks to such cooperation as you and your fellow musicians have given.
> Faithfully yours, Franklin D. Roosevelt[6]

---

[6] FDRL, President's Personal File. 1943.

Swing era dance bands became ever more popular at the president's birthday balls after the March of Dimes was founded. After FDR's death and the end of the war, the birthday galas were discontinued, and, in a simultaneous development, the economic viability of the big bands collapsed. However, the successful popularity of the "charity as show business" phenomenon continued apace for years afterward. Jazz, as the music of freedom, became even more firmly enlisted to support the polio mission. To ensure the pledge that March of Dimes-sponsored polio care would be available to all without regard to "age, race, color, or creed," the Foundation in 1943 had hired Charles Bynum, a civil rights activist who cultivated an active network of progressive African-Americans. As Director of Interracial Activities, Bynum enlisted black professionals to help fight polio, ensuring that March of Dimes programs were applied equally and fairly, especially in the South where segregated facilities, including hospitals, were legally enforced. Among the entertainment celebrities who supported the March of Dimes, jazz musicians stepped forward repeatedly. These included Teddy Wilson and Lionel Hampton, the premier pianist and vibraphonist of the age, who had broken the color barrier in jazz in Benny Goodman's trio and quartet. Teddy Wilson has been called "the Jackie Robinson of jazz" for his courageous stance with Goodman for an integrated bandstand, and it is no surprise that both he *and* Jackie Robinson supported the NFIP in appreciation of its concerted effort to maximize visible support to African-Americans, and for its pledge and actions of inclusion and equity. Over the years, Ella Fitzgerald, Duke Ellington, Sarah Vaughan, Hazel Scott, Sammy Davis, Jr., Nat King Cole, Lena Horne, and many other jazz and pop artists lent their endorsement to the March of Dimes in the movement to eradicate polio in the United States. The iconic photograph of Louis Armstrong squeezing out a high note from his trumpet aside the stage banner "*Join the March of Dimes*" may be the pinnacle of this general endorsement. Armstrong connected with the March of Dimes at the time he starred in the movie *The Five Pennies* (1959) about bandleader Red Nichols (whose daughter had polio) and his band "The Five Pennies."[7]

Despite personal endorsements from celebrities in every field of artistic endeavor, the most lucrative strategy for the NFIP to tap into the financial power of Hollywood came from the highest ranks of the movie industry. Basil O'Connor quickly grew to depend on this support from its

[7] Thanks to Phil Schaap of radio station WKCR of Columbia University at 89.9 FM for his brilliant scholarship and radio commentary in charting the history of jazz, which provides the general background to this subject here.

beginning, perceiving that he might nurture a real windfall from the rapport that existed between Hollywood and the president. Eddie Cantor had given birth to the phrase "March of Dimes" in 1937 in the very company of movie moguls of MGM Studios, but in the end this was only a convenient circumstance, that is, luck. What really mattered was the maneuvering of President Roosevelt, who has been called "the consummate media politician of his day," with Hollywood filmmakers to produce films that might steer popular opinion from isolationist pressures and set the interventionist tone in the developing global confrontation. Further, FDR pressured the studios to develop the cinematic propaganda needed for the war effort, especially after Germany and Italy banned American movies to foreclose their influence from their countries. From this came films such as *Eyes of the Navy* (1940), a prewar recruitment film from MGM explicitly touting American defense policy.[8] A year after the founding of the NFIP, in 1939, the ruling interests of the movie industry were lined up for general support of Roosevelt and his new polio foundation, and some level of endorsement was given by all the major industry leaders: Columbia Pictures, Metro-Goldwyn-Mayer (MGM), Paramount, RKO-Radio Pictures, Twentieth Century-Fox, United Artists, Universal Pictures, and Warner Brothers. After the United States entered the war in 1941, one of these—MGM—elevated its contribution by an order of magnitude primarily through the efforts of film executive Nicholas M. Schenck (1881–1969).

One might easily add the name of Schenck to the trio of Roosevelt, O'Connor, and Cantor as most responsible for popularizing the March of Dimes during wartime; unlike them, he was not generally known to the public at large. With Schenck's influence, movie theater managers from coast to coast were permitted, even mandated, to organize fund-raising in theaters just before the screening of feature films to collect contributions to the NFIP during its annual March of Dimes campaigns. These collections began sporadically in 1941 and became coordinated industry-wide in the following years. Ernest Emerling, Vice President of Loew's Theaters (the parent of MGM) who became a publicity director of its March of Dimes drive, summarized the ingredients of a successful campaign, illustrating his own fertile creativity for the March of Dimes. First, Emerling said, the local theater manager must arouse the enthusiasm of the entire theater staff and ensure there are plenty of volunteers to take collections. Attractive female

[8] Koppes, Clayton R. and Gregory D. Black. *Hollywood Goes To War: How Politics, Profits, and Propaganda Shaped World War II Movies*. Berkeley: University of California Press, 1987. pp. 33–36.

collectors were always a plus (after all, this was Hollywood). Next, the appeal should be made at *every* film showing, using open containers with lights up during the collections. The theater should advertise and feature special attractions during the March of Dimes drive, covering contiguous events in the vicinity of the theater, and place March of Dimes wishing wells and other collection devices in the theater lobby. With this prescription, the theater collections soon became wildly successful.[9] By 1944, the NFIP had streamlined the entire process to ensure maximum participation from the movie theaters: every year each state chairperson would receive a list of all theaters operating in his state, the names of NFIP county chairpersons, samples of all report forms, a complete campaign book, poster samples, March of Dimes coin cards and collection boxes, and all literature pertaining to the drive.

During the war years, the NFIP thus relied on several streams of revenue: the President's birthday ball, which had merged into the March of Dimes campaign from its origin in radio, plus market-oriented drives directed to specific constituencies: organized labor, the armed forces, sports fans, and moviegoers. Besides the several connections to the armed forces, March of Dimes ties to organized labor through the American Federation of Labor (AFL) and the Congress of Industrial Organizations (CIO) were a boon to wartime fund-raising, but voluntary payroll deduction plans sanctioned by labor and industry that began to automate the flow of individual donations from workers later turned problematic. These plans soon heralded the tendency of management to favor unified or "federated" fund-raising plans, and O'Connor's opposition to the philosophy of federated fund-raising was uncompromising, deeply rooted in his commitment to fighting a single disease from the premise of voluntary association. In 1943, however, the pledges of William Green (President, AFL) and Philip Murray (President, CIO) to the NFIP were extremely helpful, warmly welcomed, and answered directly by FDR himself with personal messages of thanks.[10] While payroll deduction plans, though not universal, began to guarantee a dependable volume of donations from the nation's workforce, the drive to engage donors during leisure hours still proved more lucrative. Radio and cinema were essential. In contrast to the overlapping complexities of the media environment today, only these two streams of "electric" media reached mass markets during the war (ignoring printed literature). Eddie Cantor's renown

[9] Emerling, Ernest. "Dimes Parade! Marches On!" *Film Daily Yearbook of Motion Pictures*, 1945, p. 134.
[10] MDA, Basil O'Connor Papers, Series 4: FDR Correspondence.

as a vaudeville star that had crossed over to radio situated him perfectly as Chairman of the "March of Dimes of the Air." Roosevelt and Cantor were masters of the radio medium, and they aided O'Connor strategically in their utilization of radio for the March of Dimes. Similarly, Nicholas Schenck entered the picture to institute an industry-wide campaign of movie theater collections giving an enormous boost to March of Dimes fund-raising for 4 years, 1942 to 1945. As president of MGM Schenck had acquired control of Marcus Loew's theater empire in 1927, and his steadfast attention to the bottom line at the helm of MGM translated into consistent profitability, even during the Depression years.

In 1943, Schenck chaired the Motion Pictures National Committee March of Dimes Drive, ensuring the industry's support from the highest level. As a result, the March of Dimes had a virtual monopoly on movie theater fund-raising as thousands of theaters were authorized, and encouraged, to permit coin collections preceding a film during the January drive. Frank Meyer of Paramount Pictures, Harry Brandt of Brandt Theatres, and Frank Whitbeck of MGM were also key supporters. Popular stars such as Judy Garland, Mickey Rooney, and Greer Garson ("the lady who found 55,000,000 hearts!") blended sentiment and civic duty in March of Dimes film trailers. In a booming wartime economy where movie theaters in large cities stayed open around the clock to cater to shift workers at munitions plants and shipyards, passing the canister between the March of Dimes trailer and the March of Time newsreel became a windfall that promised to eclipse the birthday balls and the radio appeals. To O'Connor, movie theater dollars were pure gravy, but in truth its days were numbered. Patriotism, the global war, the scourge of infantile paralysis, and the dominance of the film medium converged to ensure the popularity and success of the March of Dimes through the end of the war. The National Motion Picture Committee collected over $2 million for the 1943 drive. Nicholas Schenck and Basil O'Connor traveled to the White House to deliver a symbolic $1 million check to FDR; this was touted in a front-page photo for the *National Foundation News*. The NFIP made it abundantly clear that an equivalent $1 million had been distributed among local communities through March of Dimes chapters. In that year the totals were augmented by a British film production—Sir Cedric Hardwicke and British producers who donated the American proceeds ($240,000) of the wartime drama *Forever and a Day* (1943) to the NFIP.[11]

---

[11] *National Foundation News*, 2(8), p. 1; June 1943; 2(5), pp. 19–22; March 1943.

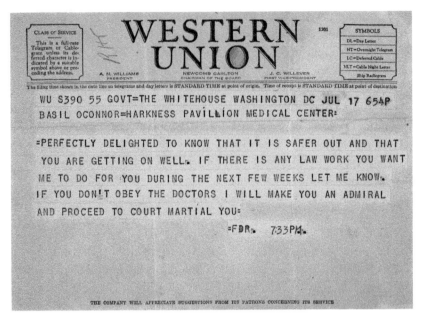

FDR Telegram to Basil O'Connor, July 17, 1942. In 1942, Basil O'Connor was hospitalized at Harkness Pavilion Medical Center in New York City, and this telegram brought him FDR's ironically humorous get-well wishes. Even in the darkest days of the war, with colossal responsibilities besetting them both, Roosevelt and O'Connor maintained the warmest camaraderie through the mutual confidence of sincere friendship.

Nicholas Schenck repeated his performance as chairperson in 1944 as Hollywood maintained its alignment with the President and the entertainment needs of a nation at war. O'Connor refused to relax his grip over this new onrush of dimes, and the NFIP orchestrated a series of pre-campaign meetings to consolidate the participation of all movie theater owners and franchises. The top leaders of the Motion Pictures National Committee—Harry Brandt, Oscar Doob, and Eddie Anderson—scheduled January meetings in Atlanta, Chicago, New Orleans, and Dallas to firm up the engagement of theater owners, producers, and distributors. The plight of those disabled by polio was vividly presented with a hands-on demonstration of the Kenny method of physical therapy. At the Atlanta meeting, Dr. Robert Bennett of the GWSF personally conducted the Kenny demo. The Dallas meeting was most enthusiastic of all, concluding with the announcement that 100% of motion picture producers, exhibitors, and owners in the 48 states had agreed to support the 1944 March of Dimes.[12] O'Connor sensed another windfall

---

[12] MDA, Fund Raising Records, Series 1: Correspondence. 1944.

on the horizon and pressed his connections not only with the President, but also with the President's men. He tapped White House press secretary Stephen Early's friend Robert Denton of Paramount Pictures whose personal efforts alone accounted for nearly 25% of the 1944 gross.[13] Even the production of the movie trailer was not beyond O'Connor's scrutiny: when the film committee substituted Greer Garson for Tyrone Power in the MGM trailer, O'Connor winced and disagreed, but insisted, "We *must* approve the narration. *That's that.*" In January, as the Greater New York chapter kicked off the motion picture drive with a rally in Times Square, movie theaters across the nation again passed the hat for the March of Dimes. When the county and state totals were calculated, the 1944 motion picture drive had reached $4,667,521, an impressive 120% increase over 1943.[14] Once again, O'Connor and Schenck paid their respects to a war-beleaguered President with a symbolic check, but O'Connor was already looking ahead to 1945. By then, as Ernest Emerling of Loew's reported:

> The activities that have been developed by imaginative showmen during the last two campaigns has resulted in several hundreds of thousands of dollars additional for the fund. The money is around—the question is how to get it. Most good showmen are exploiters and publicists at heart. They know that a "Dimes" stand at a busy corner, or in a department store, hotel lobby or railroad station will pick up many an extra dime and dollar. Coin collection boxes near the cashiers of restaurants, drug stores, bowling alleys … siphon off an amazing amount of loose change. No one has been able to prove that a store tie-up or street ballyhoo on a motion picture has actually brought extra money into the theater. We all think these things help, but with the March of Dimes we can prove they do.[15]

Not satisfied with Roosevelt's cursory *thank you* to the Hollywood moguls who had reaped $4 million from moviegoers, O'Connor sent an urgent message to FDR's secretary Grace Tully asking to rework a *thank you* letter with more lavish appreciation to the March of Dimes Committee of the Motion Picture Industry, this time in the President's own hand. O'Connor's importunate demand—"FDR *Please* do this!!" (a penciled note over his instructions to Tully)—was necessary for he well knew that FDR was burdened with overwhelming pressures and yet was the one and only person to guarantee the Hollywood revenue for another year.[16] Again the industry honored FDR with a pledge of support in the final year of the war, but this was to be the last hurrah. With FDR's death in 1945, film industry

---

[13] FDRL, Steven Early Papers, Early to O'Connor. March 20, 1944.
[14] MDA, Fund Raising Records, Series 1: Correspondence. 1943, 1944.
[15] Emerling, Ernest. op. cit. p. 134.
[16] FDRL, President's Secretary File. Basil O'Connor, 1944.

support collapsed almost entirely the very next year, and the theaters drafted a resolution that "the great majority of movie patrons have become increasingly resentful of audience collections."

Some proposed to limit the number of collections to one per year, and to open the field up to other organizations. NFIP Executive Director Joseph Savage reported that the industry saw "no more reason to collect for NFIP than for Cancer, Red Cross," and others. Appealing to the tradition of commitment to FDR became unconvincing now that the war was over and FDR gone from the scene. The sudden about-face stimulated fears that the NFIP would implode without the charisma of FDR's living presence. Before long these fears evaporated, for the reversal actually proved the stimulus to reposition the annual drive more securely in local chapters. For O'Connor, the brief romance with Hollywood ended with a kiss of death for he refused to tolerate the idea that if movie theaters could not collect for organizations separately, then "the only practical solution is one joint collection for the benefit of all." The following year he instructed NFIP state representatives and regional directors to attempt a plan for movie theater collections strictly at the local level, absent national endorsement as before. He also placed a notice in *The Hollywood Reporter* applauding the motion picture industry from his experience with its generosity to both the Red Cross and NFIP, but the wartime marriage of the movies and the March of Dimes would never be repeated at the same level of fully integrated collaboration.[17] There would be replacement activities and other opportunities for publicity, however. One was the poster child program that began officially in 1946 but had its inspiration in the disabled children depicted in the birthday ball literature. The March of Dimes posters proliferated in many markets (eg, groceries, pharmacies, restaurants, bars, and shops) through the 1950s. To approximate the spectacle of cinema, Elaine Whitelaw as NFIP Director of Women's Activities inaugurated an annual series of fashion shows in the Grand Ballroom of the Waldorf Astoria in New York City in association with Eleanor Lambert and the Couture Group of the New York Dress Institute. These annual fashion shows, which ran from 1945 to 1960, were a regular splash of publicity and glamour that included some of the most notable figures in entertainment and the arts: Jose Ferrer and Helen Hayes (Masters of Ceremonies); Inez Robb and Anita Loos (scripts); Salvador Dali, Cecil Beaton, and Alexander Calder (stage sets); George Balanchine and Tanaquil Le Clercq (performance); and Lester Gaba (stage direction).

---

[17] MDA, Fund Raising Records, Series 1: Correspondence. Motion picture campaign, 1946; memos, 1947.

Franklin Roosevelt and Basil O'Connor, 1944. This iconic photograph of President Roosevelt and his friend "Doc" counting dimes at the White House desk symbolizes the partnership of the two polio fighters and the success of the March of Dimes. By 1944, the war was approaching its final stages and the annual March of Dimes drive had become an American tradition. The perception of FDR as a "champion of freedom" was based on his principles of "Four Freedoms" and American leadership of the united nations of the free world against the Axis powers. As the embodiment of democracy and freedom, FDR's image was magnified further by his connection to the principle of Freedom from Disease advanced by the NFIP against polio.

The support of Hollywood was conspicuous, but there were others behind the scenes whose influence was just as critical. One of these was Eleanor Roosevelt. From the founding of the NFIP, Eleanor Roosevelt played a vitally important role as a goodwill agent, spokesperson for the disabled, and hostess of an annual White House luncheon just as she had as Honorary Patroness of the President's Birthday Balls. The luncheon evolved into an annual tradition, an adjunct to the birthday galas and effervescent with the sparkle of Hollywood celebrities. As the diversification of the March of Dimes drives supplanted the established regime of the President's Birthday Committee fund-raising activities multiplied, and Mrs. Roosevelt willingly lent her role as First Lady to the cause, especially with activities centered on the participation of women. Her ally in these efforts was Dorothy Ducas, NFIP Director of Public Relations from 1946 to 1962, who entered the scene as a key liaison between the First Lady and the Foundation. As an

International News reporter Ducas first became acquainted with Mrs. Roosevelt in 1934 on a trip to the Caribbean. Their friendship grew in their common commitment to social activism and subsequently to the NFIP, and they had once even contemplated a tour of the ghettoes of New York's lower East side together that seemed not far removed from an adventure in slumming.[18] Eleanor Roosevelt could claim many female friends in the profession of journalism since she had instituted her own press conferences as First Lady that were restricted to the participation of women, and she would always include Ducas among her closest friends in this field.

Ducas joined the NFIP officially in 1942 as Director of Women's Activities, a position later occupied by Elaine Whitelaw. With Mrs. Roosevelt's help, she orchestrated a social event at the White House for a March of Dimes women's auxiliary in 1938. With Ducas in charge, the "women's crusade" grew more bountiful with each succeeding year. In the autumn of 1940, Ducas organized a White House tea for the "Women's Division of the 'Fight Infantile Paralysis' Drive" for the wives of politicians in Washington. Mrs. Roosevelt served as hostess; she and Ducas encouraged attendees to organize their own teas, luncheons, bridges, and children's parties to benefit the March of Dimes. Ducas also wrote radio scripts for Mrs. Roosevelt. In 1943, when the First Lady, amid the pressures of her wartime schedule, threatened to cancel an upcoming radio panel discussion, Ducas intervened. Her personal plea paid off when she arranged a remote connection in Washington via the Blue network, apologizing for her persistence but acknowledging, "I know how much your participation means to the whole drive, which I am certain, too, is very dear to you." In their collaboration on the White House teas, which continued through the war, Eleanor Roosevelt and Dorothy Ducas cemented a friendship that lasted through the 1950s.[19] After a stint with the Office of War Information in Washington from 1942 to 1944, Ducas jubilantly informed Mrs. Roosevelt that she had accepted a part-time position "with Mr. O'Connor and my old love, the Infantile Paralysis movement."[20] Her first assignment was to engage women's magazines to advertise for volunteers for local chapters. Her forte, however, was the annual White House luncheon; in 1944, she coordinated a joint radio appeal by Mrs. Roosevelt and silent screen star Mary Pickford, then an honorary chair of the Women's

---

[18] Cook, Blanche Wiesen. *Eleanor Roosevelt: The Defining Years, 1933–1938.* New York: Penguin, 1999. p. 169.

[19] FDRL, Anna Eleanor Roosevelt Papers. Dorothy Ducas to Eleanor Roosevelt, January 10, 1943. Social Entertainments, Box 91. Dorothy Ducas correspondence and press release.

[20] Ibid, April 23, 1944.

Division. Through her friendship with Eleanor Roosevelt, Ducas grew closer to Basil O'Connor, a relationship that deepened further in the 1950s through their mutual imperative to remain in command of the media. Ducas managed O'Connor's biographical data for press releases and magazine articles, portraying him as the "crusader in the conquest of polio."[21] She even enlisted JAMA editor Morris Fishbein to assist her to recommend O'Connor for an honorary degree.[22] Ultimately, her journalistic talents culminated in forward-thinking publicity management for the NFIP during its most challenging years of battling polio epidemics across the nation. As Director of the Public Relations Department, Ducas was numbered among the so-called "matriarchy" of women under O'Connor that included Elaine Whitelaw (Director of Women's Activities), Catherine Worthingham (Director of Professional Education), May Lipton who developed health education publicity, and Maude Hussey who managed the distribution of March of Dimes campaign materials. Charles Massey, then Director of Chapters, recalled Dorothy Ducas' confident style and expertise at the height of the polio epidemics in the 1950s:

> I still remember my impression of Ducas when I attended my first staff conference at the national office. When she asked me how the Foundation's messages were being received in the field and how they might be more effective she struck me as a person almost obsessed with the desire to beat polio. As a veteran newswoman she had worked for Eleanor Roosevelt and the Democratic Party before O'Connor hired her to help jump start the foundation. She was primarily responsible for the theme of each annual fund raising campaign, including the selection and presentation of the March of Dimes poster child. Ultimately, she played a huge role in making the March of Dimes a household word. She was usually on the alert for public relations problems and ready with solutions, and when television was introduced she set up a separate division to maximize its impact. In short, it was under Ducas' leadership that the March of Dimes earned the reputation as "a powerful publicity machine," a reputation that was not always a blessing. For instance, Dr. Francis, who supervised the vaccine field study in 1954, was so wary of March of Dimes hype that he refused to issue interim progress reports. Moreover, he insisted that the public announcement of the results be handled by the University of Michigan. I remember Ducas' unhappiness at being shut out of the story of a lifetime, and I am confident it would have been handled better had she been in charge.[23]

Charlotte Jacobs, author of *Jonas Salk: A Life*, writes that Dr. Francis "practically threw out" the NFIP staff person sent by Ducas to his office to collect biographical information in advance of the field trial.[24] Ducas did not shy away

[21] *Sigma Phil Epsilon Journal*, 38(3), pp. 2–7, February 1941.
[22] MDA, Public Relations Records. Ducas to Fishbein, August 16, 1954.
[23] Massey, Charles. Interview, July 28, 2004.
[24] Jacobs, Charlotte. *Jonas Salk: A Life*. New York: Oxford University Press, 2015. p. 149.

from engaging news celebrities such as Walter Cronkite to narrate polio films or popular stars such as Lucille Ball and Desi Arnaz to endorse the March of Dimes on camera with their children in their arms. Retaining these two stars was a brilliant stroke, for *I Love Lucy* was the most watched television program in the mid-1950s; in fact, this popular sitcom was almost synonymous with the revolutionary transformation of media that television had induced. Ducas and Howard London witnessed the advent and saw the unlimited promise of television and used the new medium to every advantage. Charles Massey recalled an incident shortly after the Salk field trial announcement of April 12, 1955 in which Ducas proved her mettle as a crisis manager in an unfortunate predicament of damage control—the so-called "Cutter incident:"

> *Because manufacturing protocols were not strictly followed by the Cutter Labora-*
> *tory a batch of vaccine caused a number of cases of polio, bringing the immuniza-*
> *tion program to a halt. Vaccinations were finally resumed but the incident was a*
> *public relations nightmare for the Foundation. We were fortunate that Ducas was*
> *such a good crisis manager. Another "Ducas incident" sticks in my mind. When it*
> *was announced in 1958 that our new mission would include arthritis and congeni-*
> *tal malformations, she and her department had a visceral reaction to the term*
> *"congenital malformations." When they coined the term "birth defects," all of us*
> *were immensely grateful.[25]*

Dorothy Ducas could always rely on her friend Eleanor Roosevelt to promote the March of Dimes whenever necessary, though the First Lady's own United Nations career after the death of FDR placed her in the international limelight for the rest of her life. Eleanor Roosevelt's sympathy for the NFIP was tempered by her propensity to measure results and to criticize inefficiencies, sometimes to O'Connor's displeasure. As a First Lady whose social activism was a staple of the daily news, she was a magnet for scores of needy people who sought her assistance and intervention, including those affected by polio. In 1939, she referred to the NFIP a man who needed leg braces for his child, chiding O'Connor that the District of Columbia "seems to be doing very little" for polio cases.[26] O'Connor bristled at this charge, making sure to inform her months later when a chapter had officially been established in the district, instructing her to refer DC polio cases there. For his own convenience he reminded her to send all general polio correspondence to his office at 120 Broadway in New York City, even if it seemed apropos for Warm Springs (O'Connor shuttled between New York and Georgia, favoring his Manhattan office as headquarters). She remained businesslike in her dealings with O'Connor, always respecting his

---

[25] Massey, Charles. Personal interview. July 28, 2004.
[26] FDRL, Anna Eleanor Roosevelt Papers, Box 697, Basil O'Connor, 1939.

friendship with FDR. In 1940, she had personally invited him to a private birthday party for the President at the White House, asking him to "concoct a stunt" for the evening's entertainment and to plan to spend the night.[27]

Time and again she brought the question of racial equity at Warm Springs to O'Connor's attention. Nor did she hesitate to give her pointed opinions about unallocated chapter monies or even her preferences about the most favorable speakers for polio conferences. In 1942, she met with Marjorie Lawrence, whose operatic career had been curtailed by paralysis, and with Sister Elizabeth Kenny, the Australian nurse who in O'Connor's estimation was a problematic gadfly. (O'Connor had once counseled FDR to decline Kenny's request to allow her to dedicate her biography to him: "She writes horribly, is apt to say most anything, and for you or anyone else to let her dedicate her book to you without seeing the manuscript would be very unwise.")[28] Mrs. Roosevelt's most vociferous complaint during the war years reached O'Connor in 1945 when she intervened on behalf of Dr. George Draper concerning a discontinued NFIP grant. Dr. Draper was a family friend, FDR's attending physician in New York during his recovery from polio. She needled O'Connor that "considerable rumor is about that the advice of Dr. Fishbein is largely controlling the medical policies of the Foundation. This, of course, would be ruinous to any progress. Is this true?" "Disquieted" over Fishbein's influence, she expressed skepticism of his worthiness, which O'Connor dismissed with a resounding vote of confidence.[29] O'Connor characterized Dr. Draper's appeal as "pressure" and upheld the integrity of the NFIP medical committees, bluntly remarking, "Anyone who seeks to impugn their good faith is stupid." He needled her in return, "The National Foundation pays the penalty for not being perfect and for having been very successful. This is the kind of thing I am sure you understand."[30] He assured her, however, that the committees followed a strict protocol for awarding grants, and she relayed his assurance to Dr. Draper in a brief note. Over the years, O'Connor responded to all her queries dutifully, and their relationship warmed considerably. By the end of the war, he had changed his salutation from "Dear Mrs. Roosevelt" to "Dear Eleanor." To Eleanor he was, inevitably, "Doc."

The advent of the Salk polio vaccine in 1955 spelled the end of polio, and Eleanor Roosevelt forcefully expressed her opinion that O'Connor was most responsible for the vaccine victory. She sent a gift of $2,500 for O'Connor to transfer to Jonas Salk, earmarking it explicitly for Dr. Salk to

---

[27] FDRL, Anna Eleanor Roosevelt Papers, Box 724, Basil O'Connor, 1940.
[28] FDRL, President's Secretary File. May 20, 1943.
[29] FDRL, Anna Eleanor Roosevelt Papers. Eleanor Roosevelt to O'Connor, June 21, 1945.
[30] Ibid. O'Connor to Eleanor Roosevelt, 26 June 1945.

take a well-earned vacation.[31] On April 20, 1955, in the wake of the public enthusiasm over the Salk vaccine's licensing, she offered Doc her warmest thanks: "It seems to me they said very little about you in the broadcasts on the Salk vaccine, [but] I want to tell you how deeply I appreciate all you have done. Without your constant work and thought the Foundation would never have accomplished what it has. … I want to tell you how grateful I am for I think this is the thing above everything else that Franklin would have hoped to have accomplished." Yet, ever the conscientious guardian of morality and citizen of the world, she wrote again 2 days later on another matter entirely. Contacted by the Israeli consulate about a promise to send the new polio vaccine to Israel, she urged O'Connor either to send the vaccine or explain the situation in writing since "Israel is the hardest hit of the Eastern countries" by polio.[32] She plaintively concluded, "I'm sorry to bother you but I can understand their worry."

Eleanor Roosevelt and Basil O'Connor, 1958. Eleanor Roosevelt speaks here with Basil O'Connor in the year the NFIP changed its mission from polio to birth defects prevention. Eleanor Roosevelt supported the NFIP from its inception with birthday ball luncheons with Hollywood stars, fund-raising, radio spot announcements, and photo opportunities with children in hospital polio wards. After FDR's death, she gave generously to the March of Dimes and made an annual pilgrimage on January 30 to her husband's gravesite in Hyde Park, NY with each new March of Dimes poster child.

[31] FDRL, Anna Eleanor Roosevelt Papers. Box 3464, Basil O'Connor, May 30, 1955.
[32] Ibid. Eleanor Roosevelt to O'Connor, April 22, 1955.

Although an international political celebrity, Eleanor Roosevelt never embodied the glitz and glamour of Hollywood. Yet her own experience with her convalescent husband brought her from the onerous tedium of bedside duty as a faithful nurse to the project of Warm Springs and birth of the March of Dimes. As First Lady, she hosted an annual birthday luncheon at the White House to benefit the March of Dimes that brought her into contact with the most famous names of show business, politics, and the arts. She was indeed the celebrity who worked "behind the scenes." Over the years, she generously volunteered in March of Dimes activities and contributed thousands of her personal dollars to the Foundation. To promote polio vaccination, she joined a March of Dimes radiothon urging the public to get their Salk polio shots, and as late as 1960 she willingly acted as an intermediary to Basil O'Connor on behalf of personal appeals to her by polio patients and their families.[33] While she did not indulge in movie star glamour, her moral authority was, and still is, unparalleled, exhibited in her work for the NFIP as much in regular publicity as in the intimate advice she offered to O'Connor on behalf of individuals and nations in need. Among her world historical achievements attained in the years after FDR's death, her work on United Nations Universal Declaration of Human Rights stands out as the guarantor of FDR's "Four Freedoms" before the nations of the world. Among such accomplishments, Eleanor Roosevelt never failed to honor the memory of FDR as founder of the NFIP as she accompanied each new March of Dimes poster child every January 30 to a gravesite memorial adjacent to Springwood, FDR's childhood home in Hyde Park, New York.

[33] MDA, Public Relations Records. Roosevelt, Eleanor.

CHAPTER 7

Basil O'Connor and the American Red Cross

On July 13, 1944 President Roosevelt appointed Basil O'Connor as Chairman of the American Red Cross in the wake of the death of Norman H. Davis, a diplomat who had been Assistant Treasury Secretary under Woodrow Wilson and had served as Red Cross Chairman since 1938. The Red Cross assignment elevated O'Connor's international influence quickly and dramatically but placed a sizeable competing pressure on his day-to-day management of The National Foundation for Infantile Paralysis (NFIP). The assignment took him to many distant outposts of war in the Pacific and European theaters. In making the appointment, FDR endorsed O'Connor for his executive ability, knowledge of medicine and nursing, familiarity with "almost every part of the United States," and his leadership of the NFIP. O'Connor's statement of acceptance of the appointment was this:

> I am, of course, very honored by the President's appointment. The Chairman of the Central Committee of the American Red Cross is a position of great responsibility even in normal times. In days such as these—with all the opportunities war gives the Red Cross for service—the duties of the Chairman of its Central Committees must be overwhelming. All I can say, therefore, at this time is that I will do my best to justify the confidence the President has placed in me. In Mr. Norman Davis, my predecessor, the Red Cross had the advantage of having a man who had a deep interest in its activities and the ability to view them in the broadest way. [1]

A *New York Herald Tribune* editorial carped that O'Connor was not an impartial choice, but acceded his qualifications were "mostly cemented by good works" with FDR in the polio field.[2] In a consultation with Franklin D. Roosevelt (FDR) days before, O'Connor himself had pitched the new assignment to "the boss," believing it critical to the objectives they shared in the field of public health in the international arena, well beyond the domestic reach of the NFIP. By this juncture, both men were assured that the popularity of the March of Dimes was sustainable even in wartime, and the infrastructure of local chapters, medical committees, and fund-raising apparatus had gelled

---

[1] MDA, Public Relations Records. American Red Cross, 1938–1945. July 13, 1944.
[2] *New York Herald Tribune; Current Biography*, 1941.

Friends and Partners
ISBN 978-0-12-803597-9
http://dx.doi.org/10.1016/B978-0-12-803597-9.00007-2

and was operating effectively, with solid continuity from year to year. Could O'Connor take on the stupendous responsibility of leading the Red Cross without relinquishing his formidable posts at the NFIP and Warm Springs? There was no question but that the two men saw in this opportunity a means to broaden their approach to improving public health by coordinating these various efforts under one head, but how this was actually to be accomplished was uncertain. On the road, in the thick of a campaign tour for the March of Dimes, O'Connor dashed off a letter to FDR from Portland, Oregon warning that internal promotion from within the Red Cross, which he saw as the cronyism of selecting one of the "Washington Boys," would lead the Red Cross into chaos. O'Connor insisted, "This is another *extra-political* field where we *both again* can demonstrate how a situation should be properly handled." He added, "I'm having a wonderful trip. You have no idea what a contribution *you* have made to the whole cause of health thru your sponsorship of and leadership in this infantile paralysis movement"[3] (original emphasis).

But what actually constituted the "extra-political," as O'Connor expressed it, lay broadly across a shifting field of activities, especially in the context of the war. As a loyal Brain Trust insider, O'Connor had provided incalculable assistance to FDR as advisor and partner, but never through elected office. Yet his so-called "extra-political" posts could, and usually did, yield significant political capital to benefit the President. Their law partnership, the Georgia Warm Springs Foundation (GWSF), and the NFIP were all visible credits to Roosevelt's general influence, and O'Connor's Red Cross assignment would test him further to bring the Rooseveltian agenda onto the international stage through the Red Cross, even as it operated ostensibly as a humanitarian relief organization in the thick of war under its original constitution. O'Connor had already set a precedent at the beginning of the war, during the launch of the NFIP, to use the leverage of an established organization to help the President. That was the National Conference of Christians and Jews (NCCJ).

O'Connor's modest celebrity as the head of the NFIP in New York during the early March of Dimes campaigns was given an extra boost with his involvement with the NCCJ. The NCCJ was founded in 1927 by Supreme Court Justice Charles Evans Hughes, the Presbyterian minister Rev. Everett R. Clinchy, social activist Jane Addams, and clergyman Samuel Parkes Cadman to counteract the growth of anti-Catholic and anti-Semitic sentiment in the

---

[3] FDRL, President's Secretary File, Basil O'Connor, 1942–1944; 7 July 1944.

United States, brought to its most vile and dangerous expression by the Ku Klux Klan. Today known as the National Conference for Community and Justice, it continues its work as a "human relations organization that promotes inclusion and acceptance." From its early years the NCCJ promoted coopera- tion among different religious faiths, embraced positive race relations, and espoused "diversity," that is, valuing cultural and religious differences as a soci- etal (or corporate) asset. Yet the NCCJ was not so radical as to formulate an actual strategy for desegregation or legal restitution for injustices past and pres- ent. The NCCJ called for tolerance and understanding, and after a decade it became more vocal and more visible in the face of the Nazi menace. It cele- brated "National Brotherhood Week" on the third week in February, an occa- sion to call attention to its fight against bias, conspicuously endorsed by FDR in 1943. In the early years of the war Basil O'Connor had worked with Rev. Clinchy and Dr. Norman Vincent Peale to promote the NCCJ, even urging Eleanor Roosevelt to attend a mass rally in 1939.[4] O'Connor joined the orga- nization as head of the NCCJ Executive Committee in Connecticut for its "Mobilization for National Unity" appeal in April 1941. For Brotherhood Week in that year, O'Connor had helped to issue a public statement on behalf of the NCCJ equating the quest for national unity as support for American democracy ("the most perfect form of political community yet known") and for a "patriotism so great that one citizen cannot hate another." The NCCJ's platform centered on the acceptance of the essential contributions of every racial, ethnic, and religious group; for FDR, the unspoken subtexts to these lofty ideals were, first, a secularized creed necessarily opposed to any ideology based on racism and, second, a counter to the interference of "America First" antiwar isolationists that stymied the goal of providing aid to Great Britain dur- ing the Nazi blitz. O'Connor and the NCCJ appealed to Americans to strive toward a mental change of mind toward acceptance and tolerance, never achievable through threats of violence. As a patriotic call to uphold democracy against the encroachments of totalitarianism and war, the purpose of the New York campaign was "to arouse [Americans] against persons or groups foment- ing racial or religious hatred, and to promote friendship and cooperation among racial and religious groups." The mobilization called not for regimentation but liberty; not fear but tolerance; not suppression but open acceptance of all faiths and voluntary organizations. O'Connor emphasized that the drive was meant to attain unity, not uniformity: "Diversity means strength—and adaptability to new and changing conditions. Diversity is stimulating to progress and growth."

[4] FDRL, Anna Eleanor Roosevelt Records, Box 697, Basil O'Connor, 1939.

National Conference of Christians and Jews, October 7, 1940. Basil O'Connor headed a "mobilization for national unity membership drive" in New York City for the NCCJ. O'Connor's alignment with the NCCJ was a high-profile moment of a developing drive against intolerance and racial prejudice that figured prominently into the NFIP interpretation of polio as a civil rights issue through the 1940s and 1950s as the Foundation upheld its service pledge of polio care for all Americans without regard to "race, color, or creed." [left to right] Dr. Norman Vincent Peale, Marble Collegiate Church; Father Edward J. Walsh, President of St. John's College, Brooklyn; Basil O'Connor; Dr. Everett R. Clinchy, Director NCCJ; and Rabbi Samuel H. Golenson, Temple Emanu-El, New York City.

The implications for the NFIP of O'Connor's conspicuous solidarity with the NCCJ and Rev. Clinchy, who became a lifelong friend, was to begin to address the problem of segregation at Warm Springs. The NFIP had supported Tuskegee Institute with a grant of $1.6 million for the construction of a 36-bed infantile paralysis center, which opened in 1940 as part of the children's unit of the John A. Andrew Memorial Hospital. It was the first aftercare center for African-Americans with polio. From 1942, O'Connor served dutifully on the Tuskegee Board of Trustees, with a 25-year stint as Chairman from 1946. His participation in the NCCJ was not tangential to the goals of the NFIP but rather lent credence to its pledge to ensure the inclusion of all groups. As historian Stephen Mawdsley has affirmed, this episode "… uniquely positioned O'Connor to promote the NFIP and to speak to the important role of African Americans in the battle against polio. His celebrity and involvement with humanitarian causes conferred attention and credibility."[5] After the war, O'Connor continued to foster the message of tolerance, inclusion, and fairness, leading Mawdsley to conclude, "The mobilization of Basil O'Connor to speak on topics of tolerance and civil rights increased the interest in black Americans in the NFIP while normalizing civil rights goals among white volunteers."[6] But if his experience with the NCCJ was not a digression from his formal responsibilities with the NFIP, what divergences from the antipolio cause would the Red Cross assignment bring?

Once in the Red Cross position, O'Connor lost no time in appointing a committee to survey current operations and to develop plans for the postwar period. By early September, O'Connor formally presented to Roosevelt his review of Red Cross activities, supplies, finances, and civilian war relief. In December, he felt ready to request that FDR himself make a radio appeal for the American Red Cross War Fund.[7] O'Connor was an expert at delegation of responsibility, and his leadership of the NFIP never faltered as he led the Red Cross during the five years up to 1949. He had already studied Red Cross budget and operations intensively to manage the quantitative growth of NFIP chapters, and he directed his staff to analyze how the Red Cross apportioned its budget among local chapters as early as 1943.[8] The staffs of the two organizations independently engaged in strategic collaboration under their common leader, as when Red Cross Vice Chairman Howard Bonham and NFIP public relations staffer George LaPorte organized regular meetings beginning in 1945 to exchange "ideas, releases, radio information and other plans involving the Chairman" since public relations for O'Connor followed a "national publicity timetable."[9] A year later, at O'Connor's direction, Dorothy Ducas charged her staff to "figure out how the Red Cross solves the problem of getting more active local people involved in chapters."[10] Later, O'Connor utilized his Red Cross network to recruit key personnel for the NFIP from its ranks, and he fleshed out a strategy of humanitarian intervention that he added to the polio fight from direct experience of the aftermath of war and natural disasters. After Roosevelt's death, the annual reappointment of O'Connor to the Chair proceeded routinely through 1946 until his appointment for a final 3-year

[5] Mawdsley, Stephen. "Polio and Prejudice: Charles Hudson Bynum and the Racial Politics of the National Foundation for Infantile Paralysis." Ph.D. dissertation, University of Alberta. 2008, p. 70.

[6] Ibid., p. 77.

[7] FDRL, OF 124, 19 September 1944; 20 December 1944.

[8] MDA, Public Relations Records. American Red Cross, 1938–1945. Memorandum, February 3, 1943.

[9] Ibid., July 5, 9, and 19, 1945.

[10] Ibid., 1946. Memorandum, May 21, 1946.

term beginning in 1947. As the new Chairman, O'Connor made waves almost immediately by organizing a major restructuring of the Red Cross, abolishing its central committee of 18 members (who included FDR's close advisor Harry Hopkins and Secretary of State Edward Stettinius) and replacing it with a Board of Governors consisting of 50 members, 60% of whom were elected by chapters. With the NFIP model of grassroots chapter organization in mind, O'Connor at one stroke increased the influence of local Red Cross chapters, maximizing the democratic potential of the organization.[11] O'Connor accelerated his own learning curve by touring Red Cross European operations in October, which followed on the heels of a five-week tour of NFIP chapters in the western states. On both occasions, O'Connor insisted on grounding himself in local and regional operations through personal review and direct experience, familiarizing himself with field staff in face-to-face meetings. Stopping in London to relay a personal message from FDR to his "old friend" Lord Beaverbrook, he then met with Generals Omar Bradley, William Clark, Dwight Eisenhower, and George Patton to assess the logistics of Red Cross support of American troops and develop plans for general war relief.[12]

One problem that both the NFIP and Red Cross shared during the war was shortages of trained medical personnel, especially nurses. The publication of *Nursing Care of Patients with Infantile Paralysis* by Jessie Stevenson in 1940 was a vital first step in rallying the nursing profession to the forefront in the fight against polio, and many other supports to nursing developed over the course of the war. O'Connor supported nursing education through NFIP grants and scholarships to train and recruit nurses, and he joined the Advisory Council on Orthopedic Nursing of the National Organization for Public Health Nursing, forging a partnership that had a wide-ranging effect. This organization, with NFIP support, created the Joint Orthopedic Nursing Advisory Service (JONAS), which coordinated in-service training with epidemic relief in polio emergencies. The NFIP and American Red Cross collaborated to adopt procedures for emergency nursing service that set standards in recruitment, qualifications, terms of employment, salaries, and per diem payments of registered nurses to ensure nationwide uniformity of the treatment of nurses in polio epidemics. NFIP grants for nursing consultants in epidemics, scholarships for nursing education, and a manual for orthopedic nursing care followed. In 1946, the Foundation launched a volunteer training program called the Polio Emergency Volunteers (PEVs) to train

[11] ARCA, press release, September 1949.
[12] FDRL, President's Secretary File, 2 October 1944.

volunteer assistants to help nurses and physical therapists in the bedside care of patients with paralytic polio. Furthermore, the NFIP recruited over 10,000 nurses for local work in polio epidemics during the period 1946 to 1951, sustaining the professionalization of nursing education after the war. Midway through the conflict the two organizations mutually agreed on each organization's scope of work regarding the joint recruitment of nurses. NFIP Medical Director Donald Gudakunst directed his staff to note that the "American Red Cross assists in the care of patients during polio epidemics when a community is unable to cope adequately because of lack of funds, hospital facilities or trained personnel." Conversely, the Red Cross called upon NFIP state and regional representatives as well as state agencies affiliated with the Children's Bureau of the US Department of Labor in polio emergencies. A joint statement of responsibilities adopted by the NFIP and the Red Cross, evaluated and updated periodically, guaranteed a level of effective cooperation between the two agencies. Despite such problems as the dual recruitment of nurses, this cooperation even led to mutual support of each of their fund-raising drives, the Red Cross War Fund and the March of Dimes.[13]

As the hostilities of the war came to an end, the Red Cross was called upon to bring supplies, food, clothing, medicine, and new medical facilities to the war-ravaged countries of Europe and Asia. This winter program became the largest civilian war relief program in history, and all the allied nations were called upon for assistance. The United States had already been providing relief supplies to many European nations utterly broken with the devastation of 7 years of conflict. Based on his firsthand inspection of Red Cross operations in the chaotic aftermath of war, O'Connor focused on the need for transportation as the greatest exigency facing the European nations.[14] At his command were 3.5 million Red Cross volunteers in 3,754 chapters throughout the United States.[15] On May 31, 1945, just weeks after FDR's death and the unconditional surrender of Nazi Germany, O'Connor met Pope Pius XII at the Vatican who praised Red Cross efforts in reviving a war-torn Europe. The *Philadelphia Inquirer* soon depicted O'Connor as the "stabilizing link between our troops and home."[16] O'Connor's tour of the Pacific theater included Hawaii, Kwajalein, Guam, Saipan, Tinian, Samar,

[13] MDA, Public Relations Records. American Red Cross, 1938–1945. Memoranda, August 21, 1944; September 1, 21, and 29, 1944.

[14] MDA, Basil O'Connor Papers. Series 8: Biographical data. League of Red Cross Societies.

[15] Foreign War Relief Operations: American National Red Cross Report to the President of the United States. Washington, DC: The American National Red Cross, 1948. MDA, Public Relations Records. American Red Cross, A-62.

[16] *Philadelphia Inquirer*, 19 August 1945.

Mindoro, and Luzon, meeting with General Douglas MacArthur and Admiral Chester Nimitz just months after their conference with Roosevelt in Hawaii to plan the final blow to Japan. In this tour, O'Connor examined firsthand the logistics of supplying blood to wounded GIs through his review of mobile blood banks in the interior of Luzon in the Philippine Islands and elsewhere. He worked directly with US Army administration to improve the transport and distribution of blood supplies that proved a critical step in developing a blood bank program in the United States just a few years later. O'Connor would later admit that his personal interest in blood banks derived from his recovery from surgery 2 years earlier in which he needed 13 blood transfusions to stay alive. In 1948, he proudly pointed to the opening of the first of a nationwide series of regional blood centers in Rochester, New York as a significant Red Cross success.

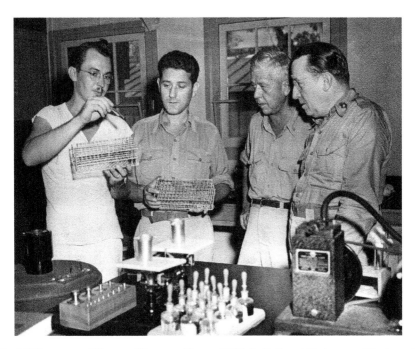

Basil O'Connor and Thomas Rivers on Guam, 1945. As National Chairman of the American Red Cross, Basil O'Connor led an inspection tour of Red Cross facilities in the Pacific theater in the spring of 1945. During the tour he reunited with his colleague Capt. Thomas Rivers stationed on Guam as head of a medical research unit for the US Naval Reserve. Here, they review the work of a virus laboratory in a Guam hospital. Dr. Rivers returned to his post as Director of the Hospital of the Rockefeller Institute and Chairman of the NFIP Committee on Virus Research after the war. [left to right] Unidentified staff pharmacist, US Navy; Lt. M. L. Hodes, USNR; Captain Thomas M. Rivers; and Basil O'Connor.

In November 1945, O'Connor was elected Chairman of the Board of Governors of the League of Red Cross Societies. The League was a federation of 63 National Red Cross Societies that had been founded immediately after World War I but which had been dormant since 1938 and the onset of the second global war.[17] With many Red Cross societies weakened and even shattered by war, the League set itself to the task of rebuilding a structure of international cooperation with O'Connor occupying a key position to leverage American interests in the process. In July 1946, he presided over an international convention of the League of Red Cross Societies at Oxford University in England, the first such conference since a prewar session of 1938. At the conference, his far-reaching proposals, which succeeded in strengthening the Red Cross and increasing its membership internationally, included a revision of the League's articles of association, establishing Red Cross societies in countries where none existed, assistance to those Red Cross Societies weakened by war, and establishing working relations with the United Nations. In a stand which seemed an extension of the Rooseveltian global, postwar vision, he upheld the need for a Fourth Humanitarian Convention, the "Protection of the Civilian Population in Time of War," which was adopted 3 years later in 1949. He also supported the proscription of atomic weapons by the Geneva Convention in addition to toxic gases and aerial warfare, which the *Red Cross Courier* optimistically reported, "may eventually outlaw the atomic bomb … through international governmental agreement."[18] O'Connor's philosophy was wholly congruent with Roosevelt's postwar vision of universal reconstruction through the United Nations, which he set out to implement through the League in tandem with the work of the NFIP and the International Poliomyelitis Conferences to follow.

The League conference of July 1946 was but one of three O'Connor attended that summer. The International Conference of Christians and Jews, also held at Oxford, passed resolutions to conduct education campaigns to promote religious and ethnic tolerance in the aftermath of Nazi racism and the Holocaust. As noted above, O'Connor had served on the executive committee of the NCCJ from 1939 and was used to speaking publically on cultural democracy and racial and religious tolerance. As a director of the NCCJ executive committee, O'Connor presented a resolution to urge national delegations attending the Paris Peace Conference to ensure guarantees of human rights in peace treaties then in development. O'Connor went on to attend a joint meeting of the National Red Cross Societies and the

---

[17] *Red Cross Courier*, December 1945, 25(6):3; *New York Times*, 16 November 1945.
[18] FDRL, Basil O'Connor Papers; *Red Cross Courier*.

International Committee of the Red Cross in Geneva for deliberations on addressing the misfortunes of sick and wounded soldiers still in the field and the treatment of prisoners of war. In his public pronouncements, O'Connor portrayed the Red Cross as "the greatest force for international peace that exists in the world today" and continually stressed its potential for promoting "compassion, humanity, and brotherhood" in its work. "The Red Cross is the only lay organization that owes its whole inspiration to a spiritual concept, the essential brotherhood of Man." Further, he insisted as a "cardinal precept" that the American Red Cross *belonged* to the American people just as he stressed uniformly in March of Dimes publicity that the NFIP was a grassroots organization founded in a concept of participatory democracy[19]:

> Very often, when I appear on the platform to speak at Red Cross chapter meetings, I have the feeling that our positions ought to be reversed—that you who represent the chapter should be on the platform and that I ought to be sitting in the audience to drink in the knowledge and experience that you have to give. We at national headquarters never forget it is the work that you *do* at the chapter level that makes the American Red Cross the helpful, successful organization that it is. Very often it has been the wisdom and the keen perception of the human need on the part of chapter members that have originated and perfected some of our most significant changes.[20]

Not surprisingly, his espousal of brotherhood, democracy, and voluntarism as an ideological formulation hinged on a stance against joint fund-raising campaigns that, O'Connor believed, diluted the identity and reduced the impact of Red Cross drives (see also Chapter 8: The March of Dimes After Franklin D. Roosevelt). He derived his strong opinion in part from FDR himself, who stated in 1942, "The character of the Red Cross and its responsibilities under International Treaty and its congressional Charter are such that the national interests will best be served if the Red Cross maintains direct contact with the people for the membership and support necessary for its work at home and abroad."[21] This was the basis of an obligation to eschew joint fund-raising because it excluded the donor from the "essential act of membership." O'Connor saw this principle enshrined in a Red Cross statement, "The Case for Freedom in Welfare Fund Raising," which confirmed his philosophy in twin principles, based on a Red Cross board resolution of April 11, 1949. The first was that "compulsory federated fund-raising, if carried to its ultimate conclusion … will serve to establish in this country a

[19] "Indispensable Corollary to World Union," *Think*, February 1947: pp. 5–6; "Opportunity and Responsibility," June 1947; Message to Delegates, June 1948; American Red Cross, annual report, 1946; introduction by Basil O'Connor, pp. 19–25.

[20] Basil O'Connor, pre-campaign tour speech, 1948.

[21] MDA, Public Relations Records. American Red Cross, printed materials, 1947–1971. Statement of the American National Red Cross Concerning Joint Fund Raising, c. 1947.

huge welfare 'trust' which is inimical to the best interests of welfare work as a whole and which is certainly contrary to American principles and practices." The second was that "the concept of federated fund-raising on a state-wide and nation-wide basis, carrying with it a dictatorial control over welfare budgets and activities, is a positive threat to Red Cross freedom of action and to the Red Cross direct membership principle which is basic to the whole philosophy and effectiveness of the Red Cross."[22] The resolution proposed that there was no popular mandate for federated fund-raising and that such systems employ coercion. It insisted that charitable giving is a voluntary act, a matter of personal faith, not to mention the statutory obligations of the Red Cross as a medium of communication between the American people and their Army and Navy in emergencies. When the personnel director of a Connecticut toy manufacturer, the A. C. Gilbert Company, wrote to O'Connor to challenge his position, suggesting that the company's fund-raising could be better served by its own Goodwill Committee, O'Connor circulated a note of commentary to his staff: "I thought this letter was going to be signed 'Joe Stalin.' … Let me know what 'you all' think before fellows like [this] prevent you from thinking!"[23]

O'Connor's leadership of the Red Cross was not without other controversies that became more public. His style of pragmatic leadership depended on indisputable control of an authoritarian, hierarchical structure and his simultaneous articulation of democracy, brotherhood, and the integrity of the individual. This philosophy bridged the NFIP and the Red Cross. When President Harry S. Truman reappointed him as Chairman in 1946, stories surfaced in the press about his cozy relationship with top labor organizations with allegations of payoffs and kickbacks during Red Cross fund-raising drives in the war years.[24] O'Connor successfully parried the charges that the Labor War Relief Committee of the CIO and AFL gained favors from his supposed malfeasance. His antagonism to bureaucratic traditionalism at the juncture of Red Cross reconversion to peacetime programs brought fundamental changes to management structure. This was far more significant, and naturally it brought forth opposition. But O'Connor's changes assured a greater fluidity of communication between the governing body and Red Cross member organizations. Equally important, the Red Cross role in global reconstruction was managed more efficiently at a time when vital services to military hospitals and supplies for prisoners of war were lifelines to civilization. In 1948, the United Nations requested that the League of

---

[22] Ibid., 1949. "The Case for Freedom in Welfare Fund Raising" 1949, pp. 3–4.

[23] Ibid. Kenneth M. Burrell, to O'Connor, June 14, 1949.

[24] *New York Herald Tribune*, 12 December 1945.

Red Cross Societies make provisions for the care of 350,000 Palestinian refugees in Jordan, Lebanon, and Syria, and O'Connor's Red Cross responded in kind. The League also repatriated 20,000 Greek children who had been separated from their families during the civil war in Greece.

O'Connor submitted his resignation to President Truman in 1949 with the statement that the position, while rewarding, was "totally consuming" and that he needed to devote more time to family and personal affairs. Upon O'Connor's resignation, President Truman appointed FDR's chief wartime military advisor George Marshall to the Red Cross post.[25] Whether O'Connor's family and personal affairs received a more significant measure of his attention is doubtful, for he quickly turned to yet another grave emergency as the NFIP struggled to cope with the largest polio epidemic ever to hit the United States.[26] However, his Red Cross experience continued to resonate within the NFIP. The next year, the NFIP sought the cooperation of the Red Cross to supply the blood fraction gamma globulin for a polio study at the Connaught Medical Research Laboratory at the University of Toronto, Canada. The Red Cross did so, and it supplied gamma globulin to Jonas Salk's laboratory at the University of Pittsburgh in 1951 as well.[27] The matter of Red Cross collaboration on gamma globulin for the NFIP was actually complicated by Basil O'Connor's dual appointment, and though his influence might have been utilized to release gamma globulin collected from Red Cross blood supplies for NFIP needs during his tenure there, a potential conflict of interest remained until his resignation from the Red Cross.[28]

Charles Massey believed that O'Connor's knowledge of logistical matters relating to blood supplies and gamma globulin "evolved more from his fertile mind than from his Red Cross experience." Massey acknowledged that O'Connor had attempted to take full control of the Red Cross, but the "entrenched bureaucrats" prevented it. Three of his Red Cross colleagues, however, followed him to the NFIP. These were Melvin Glasser, who played a leading role in planning the Salk polio vaccine field trial and the Foundation's postpolio mission; Raymond Barrows, Red Cross Vice Chairman for Health Services who served as the NFIP Executive Director; and Dr. Foard McGinnes, who coordinated relations with pharmaceutical companies during the Salk polio vaccine field trial. "All three were O'Connor's allies in his

25 *The Washington Post*, September 23, 1949.
26 Charles Hurd, *The Compact History of the American Red Cross* (New York: Hawthorne Books, 1959), pp. 251–56.
27 MDA, Public Relations Records. American Red Cross, 1950. Harry Weaver to Foard McGinnes, March 3, 1950.
28 Mawdsley, Stephen. "Fighting Polio: Selling the Gamma Globulin Field Trials, 1950–1953" Ph.D. dissertation, 2011. University of Cambridge, p. 44–45.

failed effort to change the Red Cross culture so they were happy to join him at the NFIP," said Massey, and he concluded:

O'Connor gave up his post at the Red Cross in 1949 for two reasons. On the one hand the recurring epidemics and the rapid growth of the NFIP demanded his full attention and on the other hand he had worn out his welcome at the Red Cross. … It was simply impossible for such a bureaucracy to tolerate O'Connor's one-man control, but I suspect he didn't know that when he took the job. He certainly made a valiant effort to prevail. During the war, while he was visiting Red Cross facilities worldwide, he demanded and received the prerogatives of a four-star general. I don't know whether this was a display of vanity or a display of power, but in either case I have feeling that he was just being himself. It is a popular theory that he grew up as a poor Irish guy—called "Shanty Irish" in those days—and he was determined to become one of the elite. Aside from his intellect and hard work his way of getting to the top was to dress the part and take command. While I suspect there is more than an element of truth in that theory, it takes nothing from the fact that he was an effective leader.[29]

Basil O'Connor in Manila, May 1945. Basil O'Connor examines a Japanese battle flag given to him by a GI outside the Manila Enlisted Men's Club on his inspection tour of Philippine Islands for the Red Cross. O'Connor met with high-ranking officers of the US Army and Navy to develop plans to expand Red Cross services across the Pacific after the war.

[29] Massey, Charles. Personal interview, July 22, 2004.

Basil O'Connor's swan song at his departure from the Red Cross was a keynote address "*Can the Red Cross Survive?*" presented at its 24th annual convention in Atlantic City, New Jersey on June 27, 1949. In his talk he traced the roots of the "great tradition of volunteer humanitarian relief" in the United States to the founding of the Society for the Prevention of Pauperism in New York City in 1817. He posed a rhetorical question whether a "voluntary association" was the best way for a free community to address the problem of human suffering, but his skepticism about unreformable governmental controls leading to a "monopoly of power" guided him to an inevitable answer: voluntarism is *the very bedrock* of a democratic society. Ever fond of quoting Alexis de Tocqueville's *Democracy in America*, a foundational text for O'Connor, he repeated the great French author's assertion that in the United States "the power of association has reached its uttermost development." O'Connor tolerated no contradiction with the democratic implications of that singular observation, and he stated unequivocally that the Red Cross had a special responsibility to keep alive the "spirit of free association." Upon leaving the Red Cross he continued to employ this belief with extraordinary vigor in the service of the March of Dimes.[30]

[30] MDA, Public Relations Records. American Red Cross, printed materials, 1947–1971.

CHAPTER 8

The March of Dimes After Franklin D. Roosevelt

President Franklin D. Roosevelt's (FDR) sudden death in Warm Springs on April 12, 1945 left an enormous void emotionally, psychologically, and politically in the life of the nation, and this sentiment was keenly felt in the two polio foundations he had established. James Agee wrote in *Time* magazine, "Everywhere, to almost everyone, the news came with the force of a personal shock."[1] Close associates and other observers regretted that FDR had been ironically cheated of seeing the conclusion of the European war only by a matter of weeks. By the fall of 1944, as the Allies recognized that the demise of the Nazi regime was only a matter of time, Roosevelt saw fit to return to Warm Springs for the traditional Founder's Day Dinner on Thanksgiving, an event he had missed during the onerous years of military engagement. Exhausted and ailing, he had just been elected to an unprecedented fourth term as President, and this was to be his final Thanksgiving visit to Warm Springs. On November 28, he arrived at Warm Springs for the first time since 1941, travelling from the White House to the Little White House with an entourage of close confidantes that included Basil O'Connor, as it typically did. O'Connor was toastmaster at the dinner. When actress Bette Davis crashed the party and was awarded the honor of sitting next to the President, O'Connor was furious and barely tolerated her company as he was always one to shield FDR from the importunities of Hollywood celebrities.[2] According to FDR's son James Roosevelt, by 1945 there were very few of "the old crowd" left, those with whom FDR really felt his most comfortable and informal, save for O'Connor and a few others; his daily affections had shifted to Laura Delano and Margaret Suckley, confidantes and cousins of the President. James Roosevelt spoke of O'Connor as a person who "knew and understood FDR with a rare

---

[1] Quoted in *Reporting World War II: American Journalism, 1944–1946*. New York: Library of America, 1998. p. 679.

[2] FDRL, President's Secretary File. O'Connor to Tull. October 6, 1943. See also Hassett, William D. *Off the Record with FDR*. New York: George Allen & Unwyn, 1960. p. 300.

Friends and Partners
ISBN 978-0-12-803597-9
http://dx.doi.org/10.1016/B978-0-12-803597-9.00008-4

sympathy and sensitivity."[3] Among his closest personal contacts (except-ing his physicians) James, Bill Hassett, and Basil O'Connor realized that FDR was in serious decline. When Hassett confided to FDR's physician Howard Bruenn in March 1945 that he felt that "the Boss was dying," Bruenn was concerned that the truth about FDR's carefully guarded condition would leak to the news media. Hassett admitted that the only other person who surmised the precarious condition of the President was Basil O'Connor.[4]

The Last Founder's Day Dinner, 1944. FDR makes a request to his friend Graham Jack-son, playing the accordion, at a Founder's Day (Thanksgiving) dinner at Warm Springs. This was the last Founder's Day dinner that FDR attended; he died five months later on a return trip to Warm Springs after the momentous Yalta Conference. Seated, left to right are: Charles Edwin Irwin, MD, Chief Surgeon at Warm Springs; Leighton McCarthy, Basil O'Connor, President Roosevelt, actress Bette Davis, and Vice Admiral Ross T. McIntire, personal physician of FDR in the White House and Surgeon-General of the US Navy.

An extended period of rest for FDR at Bernard Baruch's Hobcaw estate in South Carolina and then the arduous journey to the Yalta Conference to meet with Winston Churchill and Josef Stalin followed in the winter of

---

[3] Roosevelt, James and Sidney Shalett. *Affectionately, FDR: A Son's Story.* New York: Avon Books, 1959. p. 302.

[4] Bishop, Jim. *FDR's Last Year: April 1944 – April 1945.* New York: William Morrow & Co., 1974. p. 532.

1945. On March 8, O'Connor met with FDR to discuss Red Cross affairs. Afterward, on March 30, Roosevelt made the trip back to Warm Springs with O'Connor, who then returned to New York after Easter services. It was their final meeting. Though their responsibilities had separated them physically by enormous distances during the war, the two seemed spiritually closer than ever in the remaining few weeks of Roosevelt's life. FDR had given Doc an autographed copy of his "D-Day Prayer" as a Christmas gift, and O'Connor returned his thanks with the fervent wish, "I do hope this New Year will be much better for the whole world—and for you in particular that wish is meant many times over."[5] Together on their last journey to Warm Springs FDR uncharacteristically offered some private feelings toward his friend. In a moment of nostalgic reflection he actually apologized to O'Connor for burdening him with the enormous responsibilities of the National Foundation for Infantile Paralysis (NFIP), Warm Springs, and the Red Cross. O'Connor dismissed the apology, but Roosevelt persisted with this summation: "There's this to be said for your life, Doc. Most men just go down the middle of the street, doing their chosen work. You've done that with the law. But you've also gone down the sides—working for the advancement of an important science, and spending every spare hour you've got helping take care of the other fellow who's had some trouble. It's not a bad way to make the journey and I take back that apology."[6] O'Connor was prone to interpret their last moments together at Warm Springs as fraught with profound symbolism, giving him opportunities for doctrinaire reflections later on. In 1946, he reminisced: "I believe his visits to the Little White House were comforting and stabilizing experiences during his last years—the years of almost overwhelming responsibility. He could relax more completely here, and laugh more heartily here than anywhere else." He described the stark but poignant conversation of their last meeting:

> *I saw him last on Easter Sunday, 1945, a few days before his death on April 12. ... He looked pretty low. I said, "I know what your ambition is, and you are not going to realize it unless you do what I know you should do and know you won't do—take 90 solid days off." He was silent. "We'll win the war without you," I said. "Nothing can stop that. But we're going to need you much more after that. I know you won't do it, but you're on your own. Good luck." After a moment, he spoke quietly. "If I could only gain some weight," he said. To me, those were his last words.[7]*

---

[5] FDRL, President's Personal File, 96. January 16, 1945.

[6] Bishop, Jim. op. cit. p. 536.

[7] MDA, Georgia Warm Springs Foundation Records, Series 1: Correspondence. O'Connor speech, 1946.

O'Connor intuitively noted a "faraway look in his eyes" as he expressed it later.

On the evening of April 12, O'Connor, with scores of other journalists and news commentators, addressed the nation in a memorial broadcast over four radio networks, to mourn Roosevelt's passing. He reminisced about the campaign to end polio that FDR and he had initiated 20 years earlier at Warm Springs. At precisely 9:27:25 pm, O'Connor went on the air before the nation; he said:

> Today, in President Roosevelt's death, the world lost its greatest human asset. The entire universe had its attention focused upon his leadership to lead it out of the wilderness of sorrow, onto the plains of at least some happiness. ... He died at Warm Springs, Georgia, and I'm certain that if he'd had his choice he would have had it happen just that way. He loved Warm Springs, Georgia. There, twenty years ago, we started that great fight against infantile paralysis in which he was so interested. He was particularly interested in that fight—the fight to do good for the rest of humanity, to do good for those who need help and to bring some relief from their suffering.[8]

General George Marshall, Archbishop Francis Spellman, journalist William Shirer, and others touched on the many attributes of Roosevelt's greatness. Shirer reminded the radio audience that "Americans cannot forget, although they did not always realize, that it was their President who was considered by Hitler as the greatest obstacle to his evil ambitions." He talked of his recent experience in liberated France and the bond of personal friendship that so many millions had commonly felt toward FDR: "When they were liberated last summer, they felt somehow grateful to the President personally, but that was not all. Somehow, too, he represented to them the great hope of achieving lasting peace on this sorry planet."[9] In France, Albert Camus wrote in the resistance journal *Combat*:

> Roosevelt seemed to us the exemplary American. ... To be sure, we could not approve of all his policies. But whose policies can always be approved? His, at least, never bore the marks of greed or hatred. To the idealism that America has shown also has a place in reality he brought grandeur and efficiency. ... When one man succeeds to this degree, everyone succeeds. ... The greatest praise one can offer him is to say that he knew the value of life.[10]

The themes of friendship, goodwill, and moral rectitude echoed time and again. Albert Einstein opined, "For all people of good will Roosevelt's

---

[8] Geddes, Donald Porter (ed). *Franklin Delano Roosevelt: A Memorial*. NY: Pocket Books, April 1945. p. 24.

[9] Ibid. pp. 43–44.

[10] Camus, Albert and Jacqueline Lévi-Valensi (ed). *Camus at "Combat": Writing, 1944–1947*. Princeton: Princeton University Press, 2007. p. 192.

death will be felt like that of an old and dear friend."[11] Isaiah Berlin would reminisce much later, "his moral authority … had no parallel."[12] Soon after a national period of mourning and Roosevelt's interment at Hyde Park, O'Connor placed a simple, front-page tribute to the late President in the *National Foundation News* that concluded, "I can see that kindly face saying to every one of us, 'Go ahead—keep up the fight—keep going.'"[13]

There was no question but to keep going: the March of Dimes was well afloat, though there were rough seas just ahead. Some wondered whether it would flounder and sink now that it was bereft of its founder. O'Connor, above all, ensured that never happened. The spirit and vitality of Roosevelt was evoked at every turn to advance the polio mission. On some occasions, O'Connor portrayed FDR's single overriding quest as for a brotherhood of man, and he opposed these humanitarian ideals to the presence of "Hitlerism" still lurking in the world.[14] And, as attorney for the Roosevelt family, O'Connor was also ironically bound to restrain NFIP chapters from engaging in fund-raising appeals in the name of the late President himself.[15] Absent the founding father, O'Connor wrote to President Harry S. Truman in November, "For the first time in 13 years, the [March of Dimes] will get under way without the living presence of the man who mobilized the American people for war against that disease and who, during his lifetime, symbolized that fight."[16] Truman's endorsement naturally followed: "There can be no slow-down in the fight against disease." This pronouncement was tailor-made for the NFIP, cast as it was in the Rooseveltian framework of freedom from all disease, not only polio. The 1946 March of Dimes proliferated with references to the late President, radio spots repeatedly excerpted Roosevelt's speeches, and the campaign bulletin boldly asserted, "*This* war is not *yet* won."

The most enduring and tangible public symbol of FDR's intrinsic connection to the March of Dimes was the creation of the Roosevelt dime in 1946. Appropriately, the idea for the dime originated with several teenagers of a polio club who had strong ties to an NFIP chapter, not in Georgia, but in Virginia. Many of President Roosevelt's admirers were children with polio, and a substantial number were recovering in hospitals or aftercare centers. One of these was the DePaul Hospital in Norfolk, Virginia

---

[11] Einstein, Albert. Commemorative words for FDR, April 27, 1945.

[12] Berlin, Isaiah. *Personal Impressions*. Princeton: Princeton University Press, 2014. p. 26.

[13] *Red Cross Courier*, May 1945. *National Foundation News*, 4(6), April 1945.

[14] MDA, Basil O'Connor Papers. Series 1: Typescripts. Address for *Kehilath Jeshurun*. April 29, 1946.

[15] FDRL, Basil O'Connor Papers. Daniel W. De Hayes to P. C. Stone. October 28, 1945.

[16] MDA, Fund Raising Records, Series 1: Campaign Materials. O'Connor to Truman. November 1, 1945.

originally founded as the Hospital of St.Vincent de Paul in 1856, the oldest Catholic public hospital in Virginia, known today as the Bon Secours DePaul Medical Center. In 1944, in the grip of a polio epidemic that was the second largest in America to that time, the DePaul Hospital was over-burdened with paralytic children needing long-term care. The Norfolk Hospital Association chapter of the NFIP had been established in 1940 to assist with polio relief, helping to ease the burden of care with services to children and families hit by polio. The De Paul Polio Club, which consisted of polio patients all under the age of 18, was organized in 1945.

The genesis of the club was typical of the time and the setting: Rosalie B. Simmons, Executive Secretary of the Norfolk Hospital Association, overheard a boy complain of boredom after days in recuperation, and she conceived the idea of stimulating the children's recovery by forming a social activities club. The club had two purposes: to create meaningful activities for children con-fined to their beds all day and to help support the March of Dimes through creative activities. Every child in the ward, whether having polio or not, was enrolled in the club, and they elected their officers and drew up a constitution and by-laws. The group conducted weekly meetings, planned social enter-tainments, and organized March of Dimes drives. It became so successfully prominent in the hospital that two of the club's "charter members" traveled to Richmond to organize clubs at the Richmond Medical College Hospital. The officers elected were all teens under 18 years of age: Doris Bryan, Presi-dent; Richard Galloway,Vice President; Joyce Drake, Treasurer; and Richard Absalom, Secretary. Joe Brown's Radio Gang, a Norfolk-based radio program for children, often provided live entertainment, and the youngsters were gen-tly guided by parental chaperones into projects that blended competition and charity. One novel project sparked local publicity: the children fashioned little toy Scottie dogs from woolen yarn to support the March of Dimes. Eight of these woolen dogs, modeled after Fala, FDR's famous Scottish Terrier, arrived on a makeshift race course in the display window of Rice's Department Store in downtown Norfolk as a March of Dimes campaign contest. Patrons of the store placed bets on the eight numbered dogs, advancing them along a makeshift race track in the store window. First reported in the *Norfolk Pilot*, the "dog derby" was picked up by the Associated Press as a syndicated story, and the children were thrilled with the attention.

As the club's activities continued through the winter of 1945, news of Allied advances in Europe filtered in to the contained world of this tiny polio community. The death of President Roosevelt on April 12 reverberated as a shock to the world, and its impact was felt in the DePaul Polio Club as well. At a regular meeting the club members agreed that a dime with FDR's likeness would be a perfect tribute to the disabled president who founded the March of Dimes, and they passed a resolution to propose the plan to their congressman. Within the month, the club's leaders had not only reached out to Virginia's congressional representative Ralph H. Daughton with their idea, but followed up to establish the club's precedence in originating the idea in the first place, suspecting and hoping that the dime would prove significant in the continuing aura of FDR's wartime polio leadership. On November 26, Rep. Daughton introduced a bill (HR 4790) "to provide for the coinage of 10-cent pieces bearing the likeness of Franklin Delano Roosevelt," and he acknowledged the role of the DePaul Polio Club and its affiliation with the Norfolk Hospital Association chapter of the NFIP. Congress acted quickly, and the first FDR dime was minted in time for release on the anniversary of FDR's birthday, January 30, 1946. It replaced the Winged Liberty Head dime, also known as the Mercury dime, which had been in circulation since 1916. The dime was designed by John R. Sinnock though some contend that Sinnock adapted his design from the artwork of African-American sculptor Selma Hortense Burke. The *National Foundation News* noted the story with a photograph of the four officers of the DePaul Polio Club, and Doris Bryan and Richard Absalom signed and presented an official membership card to Basil O'Connor, making him a honorary member in good standing. The first Roosevelt dimes entered circulation in 1946 and have since been widely recognized as commemorating FDR's relationship to the March of Dimes.[17]

After Roosevelt's death, Eleanor Roosevelt asked O'Connor to organize the Roosevelt National Memorial committee to consider the creation of a permanent national memorial to FDR. He accepted the responsibility as temporary chairman of the committee but stepped down soon afterward with the apology that Roosevelt would have preferred him to focus his efforts on the NFIP, Georgia Warm Springs Foundation (GWSF), and the Red Cross.[18] Perhaps he was also thinking of Roosevelt's intense dislike of the word "memorial" and his instructions for the simple inscription on a plain block of marble about the size of his White House desk, which stands today at the National Archives in his memory. Over the summer, the committee surveyed suggestions for a memorial to the late President, canvassing the public and deliberating on ideas from declaring a national holiday to creating a monument along the lines of the Washington and Lincoln monuments. O'Connor was gratified with the many suggestions that keeping the March of Dimes alive was memorial enough since *that* was his own overriding personal priority. *Variety* echoed this sentiment with a long article that "[Showbiz] Leaders Would Continue

[17] MDA, Chapter Administration Records. Norfolk City and County Chapter Records, 1940–1948.
[18] FDRL, Anna Eleanor Roosevelt Records. Box 3344. Basil O'Connor, 1945–1952.

'March of Dimes' as FDR Memorial," a half-promise that would not materialize.[19] As an executor of FDR's estate, competing duties beckoned for O'Connor. He contacted Secretary of the Interior Harold Ickes, pressing him to make provisions for the maintenance of FDR's Hyde Park estate and to ensure Eleanor Roosevelt's unrestricted access. Accordingly, Ickes sent draft legislation to both houses of Congress "to protect the Roosevelt property during nonoccupancy by the life tenants."[20] FDR had drafted explicit instructions to O'Connor about the preservation (or destruction) of personal and family papers. O'Connor, of course, had already positioned himself as a loyal guardian of this documentary heritage. In 1947, he even denied access to Roosevelt's papers to Sen. Owen Brewster, head of a Senate War Investigating Committee scrutinizing military contracts to Howard Hughes. Brewster threatened to subpoena the records; O'Connor, as executor, bluntly refused him.[21]

In the summertime doldrums of Warm Springs on August 24, 1945, O'Connor presided at a ceremony to commemorate the first day of issue of the "Little White House" postage stamp, the first of several stamps to honor Roosevelt. O'Connor's reminiscence, *Nothing Could Conquer Him*, concluded, "On a beautiful Easter Sunday morning, 12 days before he died, I said goodbye to him for the last time. Here in the shadow of approaching death he was still smiling—and he left that smile here for us—nothing could conquer him!" The community of polio survivors, physiotherapists, and local folk at Warm Springs would become accustomed to O'Connor's hagiographical delineation of FDR, and there is a strong sense that he was unconsciously reaching toward a comparison with the Savior in recalling the final Easter Sunday meeting and the immortal smile of the Founder impelling the flock toward good deeds. But if O'Connor was a disciple given to speaking of Roosevelt with a messianic tinge, he was not Saint Paul—he was a practical, hard-headed, New Deal Democrat with outsized responsibilities. Although he felt the need to decline full-time chairmanship of the National Roosevelt Memorial Committee, he placed his full support behind another memorial committee. The State of Georgia created the Franklin D. Roosevelt Warm Springs Memorial Commission in 1946 to establish the Little White House as a permanent memorial to Roosevelt; Basil O'Connor was made Honorary Chairman and remained so until his death in 1972. The language of the act creating the commission also indulged in the hagiographical excesses of the day:

[19] *Variety*. April 18, 1945.
[20] NYSA, Basil O'Connor Records. Harold Ickes to O'Connor. May 25, 1945.
[21] FDRL, President's Secretary File. July 16, 1943. *New York Herald Tribune*. May 9 and June 7, 1947.

Whereas, Warm Springs, his Georgia home, is inextricably woven with the life his-
tory of the immortal Franklin D. Roosevelt; whereas it was here he found health and
contentment; it was here he rebuilt a shattered life which made it possible for him
to write his name forever on the pages of history, as well as on the hearts of all man-
kind as the greatest humanitarian the world has ever seen; whereas during the days
of the gravest crisis of all ages it was here he made so many decisions of world shak-
ing importance, decisions which resulted in the victory which saved civilization
from destruction; and whereas, when the final summons came it found him here at
The Little White House—it is but a fitting duty that the State of Georgia should per-
petuate the memory of this great man with a suitable memorial.[22]

Roosevelt had bequeathed the Little White House and its contents, as
well as his nearby farm, to the NFIP; O'Connor, in turn, deeded the property to the Commission and the state of Georgia for the memorial. The
Little White House, erected and inhabited by FDR since 1932, consequently reopened for visitors on October 28, 1948 although it had been an
unofficial tourist attraction since Roosevelt's death. Prior to opening, the
GWSF celebrated its 20th anniversary on June 25, 1947 with speeches by
O'Connor, former Secretary of the Navy Josephus Daniels, and Georgia's
Governor Melvin Thompson. Morris Fishbein advised O'Connor to use
"every available outlet" to publicize the event, including a dramatization of
FDR's illness and "his determination to establish the Foundation."[23] In
1951, with a March of Dimes wishing well installed conspicuously on the
grounds to collect donations, a memorial service was held at its chapel on
April 12. O'Connor's speech, "The Glory in the Limited Life," invoked
Roosevelt the disabled Commander-in-Chief as a role model for courageous leadership. O'Connor remained in charge of GWSF through the
1950s and 1960s though his role as president of the NFIP overshadowed all
of his other public positions after he left the American Red Cross in 1949.

His transition from the Red Cross to concentrate solely on the NFIP
coincided with the largest American polio epidemic yet experienced: 42,375
paralytic cases in 1949. The NFIP ramped up an unexpected but necessary
Emergency March of Dimes in August to attempt to replenish funds depleted by
a continuing load of aftercare exacerbated beyond capacity by a new epidemic. The Foundation arranged a benefit concert in memory of FDR on
his birthday anniversary the following year, and O'Connor's message on this
occasion was sobering: though the NFIP had aided 80% of those affected in
the recent epidemic it faced "an empty till." The concert was a magnificent

---

[22] MDA, Georgia Warm Springs Foundation Records. Series 1: Correspondence, FDR Memorial
Commission.
[23] MDA, Public Relations Records. Fishbein, Morris. February 7, 1947.

event, featuring Jean Morel conducting the Julliard Orchestra, with Marjorie Lawrence and Victor Borge as soloists and special guest Basil Rathbone who read a message of dedication from former British Prime Minister Winston Churchill, who wrote knowingly of Roosevelt's disability: "His physical affliction lay heavily upon him. It was a marvel that he bore up against it through all the years of Party controversy in his own country and through the years of world storm. As I said to the House of Commons, not one man in ten millions, stricken and crippled as he was, would have attempted to plunge into a life of physical and mental exertion, of hard and ceaseless political strife. Not one in a generation would have succeeded in becoming undisputed master of the vast and tragic scene." This profile echoed O'Connor's triumphal perspective of *Nothing Could Conquer Him* to continue to rely on Roosevelt's accomplishments to empower the March of Dimes. While the January March of Dimes campaigns of 1950 and 1951 seemed successful, O'Connor admitted again at a subsequent memorial concert for FDR in 1952 that the NFIP was $5 million in debt. Why the shortfall?

During its first 20 years, the Foundation assisted over 335,000 individuals with polio in a patient aid program that covered medical, hospital, and rehabilitation expenses. Early on, in 1942, the success of the March of Dimes was such that its revenue was greater than could be spent for research alone *at that time*. With Warm Springs in mind, O'Connor and the NFIP thus decided to attempt a patient aid program by allotting funds to chapters for patient aid at the local level. The basic formula for the allocation of funds provided that the local chapter kept 50% of the net proceeds of each annual March of Dimes campaign (after expenses) and sent 50% of the net to the national headquarters. This 50/50 formula was maintained through the early 1950s. With the 1949 epidemic and afterward, relief efforts combined with the cost of patient aid cases that carried over, sometimes for years, became so expensive that pooled funds at the national level were exhausted. The NFIP resorted to the aforementioned summertime emergency campaigns—the *Emergency March of Dimes*—to replace depleted funds in 1949 and 1954 (one was planned but called off in 1950). In this situation, with polio cases ever mounting and the Salk vaccine in development, the Foundation strongly preferred to focus on a preventative that would be more cost-effective than supporting increasing numbers of polio patients directly. By the time of the Salk polio vaccine field trial in 1954, the NFIP changed the allocation to 37.5% (chapter)/62.5% (national), which was maintained after the Salk vaccine was licensed through the period of mass vaccination programs through 1958. For a 20-year period beginning in 1942, the NFIP used this calculation to provide direct financial assistance. Individuals and families affected by polio could apply to local March of Dimes chapters for assistance, but in many cases the highly motivated chapter staff sought out polio victims when an epidemic struck. Such proactive charity was appreciated by many as the consequence of FDR's humanitarian wisdom, not only by individuals and families but by entire communities. This, in turn, boosted the prestige of the NFIP enormously and stimulated legions of volunteers to join the fight against polio for the March of Dimes.

Basil O'Connor with Schofield Siblings, 1951. The March of Dimes rarely missed an opportunity to publicize the severity of paralytic polio. The Schofield siblings of Newark, NJ were all stricken with polio, were hospitalized, and recovered—the highest incidence of siblings affected in a single family in the American polio epidemics. The five children formed a competition to raise money for the March of Dimes and are here turning over their iron lung coin collectors full of dimes to Basil O'Connor.

In the early 1940s, chapters often had the authority, provided the funds were available, to pay the entire cost of an individual's medical and hospital care for polio, regardless of a family's ability to pay. By 1947, with rising hospital costs and the soaring number of polio cases, the rationale was changed to a need-basis only. The Foundation began then to cover polio expenses after a family's resources and health insurance coverage were applied. Covered polio expenses included payment for physician services, inpatient and outpatient hospital costs, the transport of polio patients (local transport to hospitals and the Military Air Transport Service, or MATS, organized through the US Air Force), home care (including nursing care), orthopedic devices, and home medical equipment. Rehabilitation expenses often involved long-term home care and carried over from year to year beyond the addition of new cases. The NFIP philosophy on patient care (1948) was: "To provide assistance to every polio victim to enable him to receive the best available medical care without a substantial reduction in his standard of living." The Foundation claimed that it was "the only voluntary health agency in the US which conducts a nationwide program of financial assistance to patients in meeting the costs of care of a single disease." By 1956, the March of Dimes had spent $231,000,000 to help 311,000 polio patients meet the cost of their care. The polio vaccine developed by Jonas Salk, understood almost universally as an essential preventative, was appreciated by O'Connor and the NFIP in light of this burgeoning case load and the long-standing commitment to aid individuals. These numbers, added to the ubiquitous March of Dimes publicity which had entered American pop culture in the form of poster children and celebrity endorsements, ensured the popularity of the Foundation well after its transition to birth defects prevention in 1958.

To fund this program was an enormous charge. Basil O'Connor is credited for having raised "more publicity-contributed money" than any individual in the 20th century, and the totals are impressive: $3,364,000 for GWSF (1934–37); $203,000,000 for the NFIP (1938–51); and $569,790,000 for the Red Cross (1945–49).[24] In achieving these staggering totals, O'Connor adhered adamantly to the principle that fund-raising for a single cause was more effective than "federated fund-raising" by an umbrella organization. His position on this was formulated during the birthday balls and early March of Dimes campaigns with competing pressure from community welfare chests and streamlined payroll deduction plans of some labor unions that clashed with the independence of the March of Dimes. He resented the actions of watchdog organizations such as the National Information Bureau that omitted the NFIP entirely from its "Giver's Guide" in 1945, falsely suggesting that it might be ineffective, dishonest, or poorly organized. He had spoken out forcefully against the Gunn–Platt report on *Voluntary Health Agencies* in 1945 which favored consolidating and pooling public fund-raising campaigns, and his stance opposing joint solicitations or united funds supported the self-determination of NFIP fund-raising through the March of Dimes from its earliest years. To the idea that the National Health Council might be a springboard to the creation of a giant health trust he wrote, "I'm still not totally sold on the *practicality* of this kind

[24] MDA, Basil O'Connor Papers. Series 8: Biographical Data. January 7, 1952.

of an organization. To be practical I think it must control our activities to some extent. Control to some extent can lead to total control."[25] Other independent public health organizations had begun to emulate the model of the NFIP, adding to the growing complexities of the fund-raising environment. O'Connor even sought out a consultant for advice on the federated fund-raising headache. In 1949, the NFIP retained the firm of Robert Keith Leavitt of Scarsdale, New York for studies of independent versus federated fund-raising, and O'Connor mailed out Leavitt's book, *Common Sense about Fund Raising*, as a gift of enlightened propaganda to all and sundry to proselytize his position. His antagonism toward joint fund-raising was firmly rooted in his original conception of the March of Dimes as a voluntary organization based on the ideal of participatory democracy for a single cause. In his view, any united appeal weakened the impulse of volunteering for a mission in which one truly believed. Very often, the inefficiencies of federated fund-raising (and *not* the March of Dimes or its mission) were the sole subject of his speeches before service, religious, and community organizations as the competitive threat of federated fund-raising simmered through the early 1950s in the financially precarious period of the most devastating epidemics. For the lay public the Foundation even produced a primer in the form of a brochure explaining its position: "A Separate *March of Dimes* Is the Best Way to Lick Polio."

In this context, it is important to recall the direct aid to communities besides personal aid that might have been beyond the scope of personal appreciation by individual recipients. From the outset the NFIP had organized systematic epidemic relief in concert with public health authorities involving the dispatch of physicians, nurses, and epidemiologists to affected areas, with medical equipment and supplies, to cope with local polio epidemics. It had supported the helping professions intrinsically involved with aftercare: nursing, physical therapy, and occupational therapy. Above all, NFIP research grants covered not only research on polio but also on virology and basic science and resulted in funding the work of five Nobel Prize–winning scientists in the 1950s: John Enders, Frederick Robbins, Thomas Weller, Max Delbruck, and James Watson. The accomplishments of these (and other) Nobel Prize winners under NFIP grants have been reviewed in ample detail elsewhere. To facilitate communications among scientists and to pool the developing knowledge about polio as rapidly as possible, O'Connor also established the International Poliomyelitis Congress, which held international conferences on polio at 3-year intervals beginning in 1948. Aided by grants from the NFIP, the Congress went on to hold international

25 MDA, Public Relations Records. National Health Council, 1948.

conferences on birth defects after the NFIP changed its mission in 1958. O'Connor's international perspective on disease prevention, undoubtedly influenced by his work with the Red Cross, is fully evident in the methodical planning of the coordination of scientific knowledge about polio to stimulate further a holistic attack on polio in the most comprehensive Euro-American arena. The congresses also exposed Basil O'Connor's lavish hospitality to an even wider renown, for the affairs, at least initially, involved scientific presentation mixed with both high ceremony and organized socializing after-hours by conference attendees and their spouses, symbolic gifts (engraved coasters and presentation medals for foreign delegates), and associated cultural education. The NFIP scheduled the first conferences in New York City (1948); Copenhagen, Denmark (1951); and Rome, Italy (1954) in the context of the developing drive by NFIP-funded researchers to reach the ultimate solution to polio through an effective vaccine. This came in 1954 with the killed-virus polio vaccine developed by Dr. Jonas Salk at the University of Pittsburgh.[26]

Niels Bohr and Basil O'Connor, September 2, 1951. Basil O'Connor presents a memorial gavel to Nobel Prize–winning physicist Niels Bohr who presided at the second International Poliomyelitis Conference in Copenhagen, Denmark. The conference was organized under the auspices of the International Poliomyelitis Congress and was held at the Ceremonial Hall of the University of Copenhagen with Queen Ingrid of Denmark in attendance. The Roosevelt Memorial Gavel had been used during the first International Polio Conference in New York in 1948. On his return to the United States from this conference, Basil O'Connor met the young doctor Jonas Salk on board the Queen Mary.

[26] MDA, Conferences and Meetings Records. Series 5: International Conferences.

The 1954 field trial to test the effectiveness of the Salk polio vaccine has been touted as the largest peacetime mobilization of volunteers in American history. It was an unprecedented clinical trial conducted in a human population, and surely one never to be repeated again in a human population at the direction of a private nonprofit organization. With the Salk field trial, the work of the NFIP culminated after 16 years of intensive efforts in funding research to find a means to eradicate polio. Initiated, organized, and funded by the NFIP, the field trial was guided and evaluated by the Poliomyelitis Vaccine Evaluation Center at the University of Michigan established at the request of the NFIP as an impartial, independent organization. The trial involved the participation of more than 1,800,000 schoolchildren, and it confirmed the effectiveness of a vaccine that ultimately reduced the incidence of polio by 96% when put in use from 1955 to 1961. On April 12, 1955 Dr. Thomas Francis, Jr., Director of the Poliomyelitis Vaccine Evaluation Center, issued a public announcement at the University of Michigan that the Salk vaccine was "safe, effective, and potent." The announcement generated front-page news headlines in the United States and around the world, with massive and spontaneous public jubilation that the vaccine would prove to be a preventative for infectious poliomyelitis. On the same day, Oveta Culp Hobby, US Secretary of Health, Education, and Welfare, licensed six manufacturers to produce the Salk vaccine for general use. What is overlooked and deserving future study is the subsequent effort by the NFIP simply *to get all Americans vaccinated*. From its advocacy for the Poliomyelitis Vaccination Assistance Act (PL 84–377) before Congress in 1955 to authorize $30 million for grants-in-aid to the states for purchase of vaccine to the many local vaccination clinics and "polio prevention days" in the years up to the licensing of the Sabin vaccine in 1962, NFIP staff patiently endured a second battle against political snarls, public apathy, and in some cases the resurgence of polio. The NFIP public education campaign to stop polio with vaccine continued unabated from the licensing of the Salk vaccine through the time that the Foundation prepared for its new mission. As with Hollywood support during the war, the top celebrities of the day stepped up to the plate to urge the public to get vaccinated; these included Elvis Presley ("*You Can't Rock 'n Roll with Polio*") and Marilyn Monroe ("*Get Your Polio Shots Now and Play Safe*").

The field trial occurred at the height of the Cold War with the Soviet Union, just after the Korean conflict, in the mire of McCarthy-era political paranoia, and during the same year that saw the US Supreme Court decision on *Brown v. Board of Education* that began the tortuous path to the desegregation of public schools. To the credit of the NFIP and its many volunteer collaborators, the field trial ensured the inclusion of African-American children as well as those of other races and ethnicities, abetting the recent Supreme Court decision to remove the barriers of race in the burgeoning movement toward civil rights. Thurgood Marshall, who argued the case on behalf of the National Association for the Advancement of Colored People (NAACP), and later to become an Associate Justice of the Supreme Court himself, supported the NFIP and its vaccination message in a photo opportunity receiving his polio shot (along with his family) in 1957. The vociferous publicity over the benefit of Jonas Salk's vaccine counteracted, to some small degree, the fearsome image of postwar science with its hydrogen bombs and apocalyptic scenarios. The field trial announcement occurred, to the day, on the 10th anniversary of FDR's death, and some accused O'Connor of grandstanding, using the anniversary as a political symbol. O'Connor denied that the date was deliberately selected to coincide with this anniversary (and it was not), but the significance of the date in bringing FDR again to the forefront was lost on no one. The letters of gratitude and congratulations that poured into Jonas Salk after the announcement, along with a mountain of unsolicited gifts, ranged from the monumental to the absurd. O'Connor, too, received a flood of private congratulations, largely letters from friends and associates who recalled FDR's personal struggle to defeat polio. One suggested that April 12 be named a national holiday, and others chastised the news media for its lapses in forgetting to mention FDR, lamenting that "there was no adequate statement or showing of feeling for the founder of the National Foundation" or his "sustained leadership" in the fight to defeat polio. William Hassett complimented O'Connor "how fitting it was to make it on the tenth anniversary" of FDR's passing, commenting wistfully, "The circle of friends who knew and loved FDR as you did, is drawing in. That is another reason why I cherish your friendship the more as the years pass." Everett Clinchy, recently of the National Conference of Christians and Jews but now of the World Brotherhood in the 1950s, made this personal appeal to Doc: "Now you should turn your machinery to developing a vaccine against bigotry in the world. We'd love to have your help." A former Red Cross associate admitted, "I will never forget what a tough guy you sometimes were to work with, but also I will never forget the energy, and effort, and personal devotion you have given to bettering the welfare of so much of our population." "A message from the American Cancer Society was definitive: 'This country will have been a better place because you have lived.'"

Albert Sabin and Jonas Salk, 1954. The Third International Poliomyelitis Conference was held in Rome, Italy sponsored by the NFIP, the Orthopedic Clinic of the University of Rome, and the government of Italy. Pope Pius XII gave a special address, and Drs. Sabin and Salk delivered papers on their current research for the panel on Infection and Immunity in Poliomyelitis. This moment of relaxed conversation belies their fractious relationship. Basil O'Connor is seated directly behind the two famous NFIP grantees.

The history of the vaccines developed by Jonas Salk and Albert Sabin as an American success story has been revisited from several different perspectives, but in the thick of the organized fight to end polio, Basil O'Connor's own participation was made frighteningly personal: his own family was stricken by polio. O'Connor's daughter, Bettyann Culver, contracted polio in 1950, famously reporting to her father, "Daddy, I've gotten some of your disease." O'Connor elaborated on her experience in a *Look Magazine* article that featured a photo of Bettyann with her parents and FDR (supported by a cane) in 1931.[27] O'Connor's nephew, Dr. Harrison O'Connor, had also been stricken, and recovered, from the disease. Bettyann spent a period of convalescence at Warm Springs under the care of Dr. Robert Bennett. By this time, O'Connor was as familiar with polio as one might conceivably be, and this personal tragedy simply augmented his determination to see its elimination. When Walter Winchell had the audacity to suggest on national radio just prior to the field trial that the Salk vaccine was contaminated with

[27] *Look Magazine.*

poliovirus, O'Connor went apoplectic that a journalist who built his reputation on gossip and sensationalism might foil the NFIP's careful planning. He was so incensed that he rebutted Winchell on national television, speaking not only as the chief of the NFIP but as *a parent to parents* in order to allay fears that the Salk vaccine would be anything but helpful. He surely smiled at the suggestion offered in one of the congratulatory letters received after the field trial announcement that he should "send Walter Winchell a nice fat crow."[28] O'Connor was also profiled in *Good Housekeeping* magazine in July 1953 in an article by Alice Beaton, "A Friend, and Partner." His daughter's experience with polio received slight attention in favor of portraying the now famous partnership that had launched the fight against polio: Roosevelt and O'Connor. Beaton portrayed O'Connor a hard-headed businessman rather than a sentimental humanitarian, quoting him saying, "Polio is a phase of the nation's business—and for adequate fund raising to provide care and research, it required a business set-up that was nation-wide."[29] Helen Hayes was also quoted, preferring to emphasize O'Connor's life of volunteerism.

Turnley Walker's memoir, *Roosevelt and the Warm Springs Story* (1953) proved the first in a series of books and films that permitted a glimpse at the personal hardship of FDR's polio disability and his association with Basil O'Connor. Walker's own convalescence at Warm Springs stimulated an intense admiration for its founder, and O'Connor praised Walker's book as "the first detailed history of the Georgia Warm Springs Foundation."[30] This was not surprising since not only O'Connor but his Dartmouth friend Henry Urion along with Grace Tully, Dorothy Ducas, Margaret Suckley, and Frances Perkins advised Walker in his research. In a foreword, Eleanor Roosevelt stressed that the book "caught the feeling of the Warm Springs community." In an unpublished memoir, "Polio Diary," Walker described the claustrophobia of lying in bed at the Hospital for Special Surgery in Manhattan just as the cornerstone for the United Nations is set into place, fantasizing a visit from President Truman. He wistfully reflects, "There used to be a President who knew something about Polio, but he is no longer with us." Walker's wife then appears with a letter that gives them both hope for recovery and for the future:

> She reads it to you, and gratefulness at what it says made you weep together. There is an organization called the National Foundation for Infantile Paralysis ... Oh yes, you had heard of it ... something called the March of Dimes to which you gave a few dollars every year ... but the important aspects of it have escaped your attention.

[28] MDA, Salk and Sabin Polio Vaccine Records. Series 1: Correspondence. Congratulatory letters on Salk vaccine report, 1955.

[29] Beaton, Alice. "A Friend, – And Partner." *Good Housekeeping*. July 1953, p. 209.

[30] MDA, Public Relations Records. Walker, Turnley.

The letter says that you are not to worry if you do not have the money to pay for your treatments. The letter says that the National Foundation for Infantile Paralysis will stand behind you, taking care of all your medical and hospital expenses if need be, until you are able to go to work again.

Until you are able to work again, the letter says. It lets you know that you were not lost forever, but only out of circulation for a while because of a dreadful accident. … After she is gone, you can still see her smiling, and you see that the letter means precisely what it says. This Foundation has suddenly become a personal and powerful friend of yours.

Someday, you decide, the full story of the Foundation's work will be told so truly and so clearly that knowledge of it will join the traditions of the country, which are passed on in simple language from one generation to the next. Now you think of the personal kindnesses of the Foundation people you have met.

And, with these people you think of a man as yet unknown to you, the close friend of President Roosevelt in whose name the Foundation was begun, the man who made secondary a distinguished personal career, to encompass with his mind and heart the whole fight against this terrible disease—Basil O'Connor, who directs the entire activity, without thought of recompense, personalizing the amazing generosity of the organization.[31]

The cloying emotionalism of Walker's characterization seems passé today, but O'Connor gladly suffered even such maudlin adulation as this whenever his name was conjoined to Roosevelt's. When news broke in October 1953 that a polio vaccine was in development, Walker wrote promptly to his agent that "Basil O'Connor is the real hero of this accomplishment. He is the man who took the largely emotional concept of FDR and shaped it into the most remarkable, and effective, organization of its kind in the history of the world."[32] But what exactly was the "emotional concept of FDR"? In building the NFIP, O'Connor had parlayed the story of a fallen aristocrat's rise from polio paralysis to become the catalytic force behind a voluntary organization, itself a mirror of democracy. In so doing he united faith in and of ordinary people (volunteers) to the noble aspirations of progress in medical science (knowledge) via a simple common denominator of exchange (the dime). As myth-maker, O'Connor translated the image of Roosevelt's struggle into identification with a cause in which all of humankind would be beneficiary. The NFIP radically transposed the story of Roosevelt's disability into the visual imagery of the poster child as a symbol of hope and an appeal for funds: Walker recognized this implicitly as an instance of emotional branding, a strategy implicit in NFIP publicity that had proven enormously successful. Following the release of his book, Walker appealed to

---

[31] MDA, Public Relations Records, Turnley Walker. Polio diary.
[32] Ibid., Walker to Constance Smith, October 8, 1952.

O'Connor to endorse retrospective treatments of his personal story. He even contemplated a movie version of *Roosevelt and the Warm Springs Story* casting O'Connor as the central character and despite the Roosevelt family's commitment as advisors for a concurrent movie production of Dore Schary *Sunrise Over Campobello*. In that film, Louis Howe took center stage as FDR's loyal lieutenant, and Greer Garson, whose March of Dimes appeal in *The Miracle of Hickory* (1944) was a special moment in NFIP filmmaking, played Eleanor Roosevelt. In 1964, Walker wrote again to O'Connor to pitch a concept for a book that would tell the story of the Foundation's birth defects mission as one analogous to polio. O'Connor looked favorably on the movie idea (though nothing came of it), but his reaction to the proposed book was lost in his own workaholic lifestyle: he was simply too busy.

As a witness to decisions of world historical importance from the perspective of the White House and his own executive positions, O'Connor had a keen sense of the management of historiography (despite Dorothy Ducas' claim that he was inept at "selling himself"). As Jonas Salk's vaccine loomed as the answer to polio, O'Connor made plans to preserve the history of the NFIP in meticulous detail in hopeful anticipation of the vaccine triumph. Early in 1953, he retained Hackemann & Associates, a management consulting organization based in Madison, Wisconsin to direct a project documenting the history of the NFIP. Louis Hackemann, president of the firm, had researched a history of the Red Cross authorized by O'Connor; in his opinion, the field trial had been "a significant contribution to the social history of the world."[33] With Hackemann, the NFIP established a temporary "Historical Division" of writers to produce a professional history of the Foundation. He and his assistant Ruth Walrad conducted interviews of the essential players associated with GWSF and NFIP, and their writers reviewed documents in the NFIP records center to produce the multivolume *History of the National Foundation for Infantile Paralysis* in 1957. The team of writers included Christopher Lasch, who rose to prominence as a historian and social critic with his best-seller, *The Culture of Narcissism* (1979). The final document in this historical project was the compilation of a series of monographs on every phase of the work of the NFIP totaling almost 3,000 pages. NFIP Executive Director Raymond Barrows and Dorothy Ducas were liaisons to Hackemann for the duration of the project. Barrows and O'Connor expected that an

---

[33] MDA, Salk and Sabin Polio Vaccine Records. Series 1: Correspondence. Congratulatory letters on Salk vaccine report, 1955.

"eminent historical writer" would be located to condense the massive document into a publishable form, but after a fruitless search a writer never materialized.[34]

One of the researchers on the project was Adam Lunoe, an NFIP public relations staffer who had convalesced at Warm Springs and helped Turnley Walker with research on another book, *Rise Up and Walk*. Lunoe revealed to Barrows that in working with Walker, the two had "uncovered great quantities of documents" about O'Connor and Warm Springs. Walker had tape-recorded many interviews with people key to the polio story. Of the recordings then still in Walker's possession, Lunoe estimated the extent of the interviews in excess of one million words, and he asked permission of Barrows to tap into this source by visiting Walker in California. He next sent a request to O'Connor himself: "This is to ask your permission for me to examine the F[ranklin]–B[asil] file housed in a closet in your office on the 30th floor. I understand that this file contains correspondence and other material between you and President Roosevelt through the years of your association, including basic material dealing with Warm Springs affairs as well as [the NFIP]."[35] Lunoe's request fell on O'Connor's utterly deaf ears, for the most revelatory communications between O'Connor and FDR were at stake, and Barrows took the matter up in a letter to Hackemann:

> Mr. O'Connor was adamant about Lunoe's [not] going into correspondence contained in his personal file from F.D.R. and others, also because of this I refused to make available to Adam another file of correspondence between F.D.R. and Mr. O'Connor on the belief, after inquiry, that these files do not have any specific bearing on the development of Warm Springs or the NFIP.[36]

Hackemann tiptoed around this issue of access and confidentiality in the matter of O'Connor's personal correspondence, confiding to Barrows that lay people underestimate the importance of source documents in writing history. He explained that O'Connor had once helped him pry loose a set of records from a Red Cross executive in writing the Red Cross history. O'Connor, on the other hand, was not about to reveal the intimacies of his privileged relationship with FDR, though Hackemann ultimately gained access to some portion of the O'Connor & Farber legal records as well as Walker's sequestered documents. The latter included "the precious book owned by Warm Springs coverning [sic] the drive for funds for Georgia

---

[34] MDA, History of The National Foundation for Infantile Paralysis Records. Raymond Barrows memo. September 28, 1953.

[35] Ibid. Lunoe to O'Connor. October 19, 1953; November 2, 1953.

[36] Ibid., Barrows to Hackemann. November 19, 1953.

Hall" and a "scurrilous" account about FDR's death.[37] The Historical Division had planned to interview the participants in the 1954 field trial, but this crucial episode failed to appear in the history. Hackemann tapped historian Philip Jordan of the University of Minnesota to convert the inelegant but overwhelming NFIP history into a marketable book, but O'Connor expressed skepticism since "Arthur Schlesinger, Jr. does *not* know [Jordan]."[38] Hackemann and O'Connor preferred Schlesinger himself (notable historians Allan Nevins, Henry Steele Commager, and Richard Shryock were also considered) but he was immersed in his own monumental study of the New Deal, *The Age of Roosevelt* (in which he pegged O'Connor as a "shrewd, salty Irishman").[39] Hackemann & Associates completed the final draft of the *History of the National Foundation for Infantile Paralysis* in 1957 to little fanfare, eclipsed by the reorientation of the NFIP toward its new birth defects mission. The study remains a monumental but unwieldy institutional history, one that never found a publisher, and both the "Franklin–Basil File" and the Turnley Walker "archive" remain unaccounted for.

Like many convalescents and physiotherapists who thrived in the environment that FDR had nurtured at Warm Springs, Basil O'Connor found more there than his life's calling—he also found romance. After the death of his wife Elvira of coronary thrombosis in 1955, O'Connor married Hazel Stephens (nee Royall) on June 12, 1957. Born in Panama City, Florida in 1913, Hazel was a physical therapist and Director of Functional Physical Therapy at GWSF. She had joined the therapy center as Director of Recreation in 1943 and had coauthored the only known eyewitness account of Warm Springs on the day of Roosevelt's death. Her writing was later published in *The Georgia Historical Quarterly* in 1991. O'Connor had developed an especially warm relationship with Hazel and had corresponded with her formally as early as 1953, awarding her a $1,500 "travelship" through the NFIP to attend the first congress of the World Confederation for Physical Therapy in New York.[40] In working together with O'Connor to improve physical therapy services at Warm Springs, Hazel had developed a cordial relationship with Jonas Salk, who served as best man at the O'Connor wedding in 1957. Edward R. Murrow interviewed the two newlyweds on his television program *Person to Person* later that year, and Hazel soon assumed an active role in the NFIP. Their marriage endured through to O'Connor's death though it was fraught with the incessant demands of his commitment

[37] Ibid., Walker to Lunoe. December 1, 1953.
[38] Ibid.
[39] Arthur Schlesinger, Jr., *The Age of Roosevelt*. New York: Mariner Books, 2003.
[40] RWSIR, Correspondence. Royall, Hazel. August 6, 1953.

to the March of Dimes at a time of internal change. Hazel O'Connor also saw her husband through the tragedy of the early deaths of both his daughters; in 1961, Bettyann Culver succumbed to cancer; her sister Sheelagh died from an attack of pneumonia 5 years later. Hazel collaborated with Elaine Whitelaw to organize volunteer leaders for Mothers Marches and naturally stepped forward as "Mrs. Basil O'Connor"—a high-profile ambassador for the Foundation on formal occasions, such as the Second International Conference on Congenital Malformations in New York City in 1963. Hazel was awed by Doc's perfectionism and tickled by his sense of humor, which admittedly was not evident to everyone. As his spouse, she interpreted his persona as bifurcated by tenderness and authoritarianism: "I don't think he ever intended to frighten anybody, but no one would ever have been afraid of him if they had known him because he was the softest guy in the world actually, but he was a very sophisticated, honest, caring man." In an interview in 1986, she revealed a story about Basil O'Connor's compassionate qualities involving the rescue of two homeless waifs that few would ever expect to deduce from his standoffish superiority:

> [W]hen he was at Dartmouth, one of his classmates for whom he had great respect became the individual who was responsible for Catholic Charities in New York City. In this position he came across these two boys who lived under the Brooklyn Bridge, and I think there were about eight or ten children, but these two and one other child were the only ones that didn't have TB. Doc gave them jobs after school. Both boys, Dick and Bill Donovan, came to his law office every afternoon after school, and he trained them himself. They both stayed with him until they died—one with the law office and one was with the National Foundation [in charge of the mail room]. … I want you to know that that boy actually worshipped Doc. I think if Doc had told him to shoot me, he would have … he was so devoted to him. I remember him telling me how Doc had trained him. Little minute things were just as important as the great big things. The water in the carafe on his desk, every pencil sharpened, and a white carnation for his lapel.[41]

During the 1950s, O'Connor had warmed to a close and special relationship with Jonas Salk, a grantee of the NFIP. Today, such an intricate personal relationship between grantor and grantee might suggest a conflict of interest or at least a dangerous compromise to the ethical factors in grants administration. But this was not the case. Hazel O'Connor stated that "the relationship between the two bridged a time in Jonas' life that might have been very difficult for him otherwise," believing that her husband's encouragement in the face of hostile criticism gave Dr. Salk a vital shot of confidence.[42] Some have likened the O'Connor/Salk partnership to a

---

[41] MDA, Oral History Records. Dillmeier, Hazel. Interview, 1986.
[42] Ibid.

"father–son" relationship, in which they shared the lofty ideals of science in the service of humanity later realized in the creation of the Salk Institute for Biological Studies in 1960. O'Connor was well versed in the humanities and always reserved the right to question science from the perspective of law and morality. He believed fully in the philosophical exploration of the value of science in civilization to achieve a better society. For O'Connor, the idea of the Salk Institute extended the goal of "freedom from disease" onto a new institution, and though there was a falling off in the warmth of their relationship after the Salk Institute was founded, their collaboration recapitulated in a small way the kind of partnership that O'Connor had achieved with Roosevelt in the days of building the center at Warm Springs.

Polio Hall of Fame Dedication, January 2, 1958. The GWSF memorialized the end of polio by commissioning the sculptor Edmond Amateis to create bronze busts of those whose work led to the most significant breakthroughs. Eleven of those so honored attended the dedication ceremony of the Polio Hall of Fame in Warm Springs. [left to right] Dr. Thomas Rivers, Dr. Charles Armstrong, Dr. John Paul, Dr. Thomas Francis, Jr., Dr. Albert Sabin, Dr. Joseph Melnick, Dr. Isabel Morgan, Dr. Howard Howe, Dr. David Bodian, Dr. Jonas Salk, Eleanor Roosevelt (representing her late husband), and Basil O'Connor. Dr. John Enders, whose bust also adorns the frieze, did not attend the ceremony due to illness. The four busts at the far left represent four European scientists: Dr. Jakob Heine, Dr. Karl Medin, Dr. Ivar Wickman, and Dr. Karl Landsteiner.

The polio vaccines developed by Jonas Salk and Albert Sabin under NFIP grants effectively ended the polio epidemics in the United States. Polio cases diminished to only a handful by the early 1960s, and complete eradication was announced in 1979. The NFIP memorialized the coming end of polio in 1958 by celebrating its 20th anniversary, unveiling a "Polio Hall of Fame" at Warm Springs with busts created by the sculptor Edmond Amateis. The busts represented the scientists whose research breakthroughs led directly to the understanding and defeat of polio; there were only two nonmedical honorees: FDR and Basil O'Connor. David Bodian, John Paul, Albert Sabin, Thomas Rivers, Jonas Salk, and others attended the ceremony. Eleanor Roosevelt was there to represent her deceased husband. But with the public celebration of the defeat of polio, the NFIP had also reached a turning point: on July 22, 1958 O'Connor held a press conference at the Waldorf Astoria to announce the launch of an "Expanded Program" against disease, specifically arthritis, birth defects, and virus diseases. Heralding a vision that the NFIP might become a "flexible force" in the field of public health, the mission change occurred at a time of declining revenue, and institutional difficulties ensued as March of Dimes volunteers balked or failed to understand the necessity for change. O'Connor, Joseph Nee, Charles Massey, and their staffs began to address the massive organizational change with intensive reeducation into a completely different mission. The mission transformation had actually been years in the making, and O'Connor had placed his long-time associate from the Red Cross, Melvin Glasser, in charge of spearheading the change in 1953. The studies undertaken by Glasser included two comprehensive public opinion surveys by the American Institute of Public Opinion (Gallup Poll) and the Bureau of Applied Social Research of Columbia University (see Chapter 1: Introduction: The March of Dimes and Historiography). Both surveys validated the recognition of "March of Dimes" as a brand with excellent name recognition and integrity, helping to provide a framework for decisions about the new mission. In this process, O'Connor asked George Gallup personally to focus on two questions: "What does Roosevelt mean to you?" and "What … does Roosevelt stand for?" Most respondents answered that he was a "friend of the poor," and about 14% described him as "a leader in the fight against infantile paralysis."[43] While these survey questions may have satisfied O'Connor's undying curiosity about the public perception of the NFIP founder more than a decade after his death, it also illustrated that such questions were becoming increasingly irrelevant.

[43] FDRL, Roosevelt Memorial Commission.

In 1957, the New York Academy of Sciences elected O'Connor as a "gold member" for his "outstanding scientific achievement and promotion of science." The following year he received the Albert Lasker Scientific Award of the American Public Health Association, the first layperson to do so. He held a conspicuous position as President of the National Health Council during 1957–58, and with the defeat of polio and transformation of the March of Dimes he became something of an elder statesman in the field of public health. Ever the speech-maker, though never a charismatic speaker, O'Connor quietly assumed this role to weigh in on subjects beyond the fields of medicine and public health. He extolled the importance of education in "The Future of Leadership," claiming that his early education led him to the two key friendships of his life: with FDR and Jonas Salk. Upon receiving an honorary Doctor of Laws degree from Roosevelt University (Chicago, IL) in 1964 he recapitulated this theme in "The Heightened Requirements of Leadership."[44] At his most irascible, he lectured March of Dimes staffers in pre-campaign meetings on "Government by Elite Vigilantes," lamenting a new era of watchdog agencies like the National Information Bureau that he believed set intrusive standards in evaluating nonprofit foundations. O'Connor characterized the development of oversight agencies as a threat by "small pressure groups" to the "American way of life" and in particular questioned the role of the Rockefeller Foundation as the "keeper of the conscience" of organizations like the March of Dimes. O'Connor was wont to rail against conformity, regimentation, and the domination of bureaucracy, popular targets that became more popular as the turmoil of the 1960s escalated. As the influence of governmental agencies such as the National Institutes of Health became more pervasive, O'Connor tended to castigate the NIH as an example of a "perilous partnership" between science and government, an insidious development leading the bureaucratic drift away from volunteerism. His pronouncements made headlines and generated controversy, for he never withheld criticism when he felt it was deserved.

O'Connor continued to guard the Roosevelt legacy within the March of Dimes and without. When the Eleanor Roosevelt Cancer Foundation was established in 1959, O'Connor complained to its chairman General Omar Bradley about its appropriating the image of the Roosevelt dime in an AFL-CIO fund-raising advertisement for cancer. Coveting his brand and pointing out the irony of using a "dime" for labor's "march" against cancer

---

[44] MDA, Public Relations Records. Rockefeller University.

O'Connor insisted, "The words 'March of Dimes' have been used by [the NFIP] for at least 20 years … identified by the public with The National Foundation. … At the very worst, I am afraid your fine organization might be subjected to the criticism that it is deliberately attempting to infringe on a campaign whose genesis is historically known and whose theme has become a household word." Neither was he reluctant to suggest that Mrs. Roosevelt herself might experience embarrassment that the March of Dimes would be harmed by this "infringement."[45] O'Connor chaired the FDR Memorial Committee for the Benefit of the NFIP to celebrate FDR's birthday in 1957 with a dinner at the Waldorf Astoria and a 60-minute dramatic tribute to FDR.[46] Ten years later, this kind of tribute reached its last gasp when the March of Dimes created the Franklin D. Roosevelt Award. For 3 years, the award was presented to honor a person who "most exemplifies the ideals of Franklin D. Roosevelt." The award recipients were President Lyndon Johnson (1967), Ambassador Averell Harriman (1968), and Vice President Hubert H. Humphrey (1969), which reveals more about O'Connor's alignment with the Democratic Party than the merits of the individuals selected. After a year's hiatus, I. W. Abel, President of the United Steelworkers of America (USWA), was the intended recipient in 1971, but the award was canceled and collapsed into obscurity.[47]

Basil O'Connor led the March of Dimes into a new era, with a new mission that brought genetics and perinatology into sharp focus to penetrate the mysteries of birth defects, yet he grew increasingly out of touch with the encompassing political realities of the 1960s. One symptom of his detachment was a public controversy with Albert Sabin that unraveled in the press over the licensing of the Sabin polio vaccine; another was the revelation that he had begun to draw a salary from the Foundation that cast unneeded and unwarranted suspicions about his own decades of volunteering. Yet another was his response to a hostage situation at the Tuskegee Institute in 1968. O'Connor had been the Chairman of the Board of the Tuskegee Institute (now Tuskegee University) from 1946, forging ties with the black community through his friendship with Tuskegee President Luther Foster and through the outreach of the African-American civil rights leader Charles Bynum for the March of Dimes. O'Connor's commitment to financing nursing education and polio-related programs at Tuskegee remained strong for over two decades, and since his days with the

---

[45] MDA, Public Relations Records. Eleanor Roosevelt Institute.
[46] FDRL, Rexford Tugwell Papers. O'Connor to Tugwell. January 14, 1957.
[47] MDA, ibid. Franklin Delano Roosevelt Award.

National Conference of Christians and Jews he periodically addressed civil rights issues very pointedly, as in his speech *After Desegregation, What?* On this occasion, in the aftermath of the assassination of Martin Luther King, Jr., he declined to cancel a board meeting at Tuskegee despite the turmoil that racked the nation over Rev. King's murder. When Tuskegee students then threatened to hold hostage the entire board of trustees, the tense situation was diffused by Luther Foster and Melvin Glasser but not before O'Connor had naively suggested that the students simply leave the premises and place their grievances in writing instead. Such a forceful challenge to authority by militant students was completely outside his framework of understanding, yet once back in New York he promptly responded to their grievances, saying that Tuskegee must "render near-miraculous services in these times of gravity for our entire nation."[48] O'Connor's evolution from the reluctant humanitarian advising Roosevelt to steer away from Warm Springs to become an international figure in public health who had energized philanthropy seemed oddly frozen in the mindset of the New Deal. Yet even toward the end of his tenure as President of the March of Dimes, he sought to rise above what his own instincts seemed to hold back regarding supplications to the federal government, as when he pressured Senator Charles Percy to support the Neighborhood Health Center Act in 1968:

> Our deep concern with the delivery of health services stems from a conviction that birth defects, prematurity, and other forms of reproductive wastage are associated to a significant degree with deficits in the social organization of medical care, primarily with the failure of a large proportion of pregnant women to obtain early and adequate ante partum care.[49]

The language in this appeal ("reproductive wastage") is a Rooseveltian throwback to polio, but the message expresses the holistic approach to maternal health toward which the March of Dimes was evolving in its steps toward a thoughtful agenda of advocacy. Basil O'Connor will be forever identified with the NFIP program against polio, but here we find him advocating for the health of women and infants to broaden the base of support for prenatal care, especially for low-income women. Under O'Connor's direction the Foundation had begun to establish birth defects treatment and

---

[48] MDA, Basil O'Connor Papers. Series 7: Memoranda. April 6, 1968. Oral History Records, Melvin Glasser interview. September 12, 1983. Luther Foster interview. September 23, 1983. See also Stephen Mawdsley, "'Dancing on Eggs': Charles H. Bynum, Racial Politics, and the National Foundation for Infantile Paralysis, 1938–1954," *Bulletin of the History of Medicine*, 84 (2010): pp. 217–247.

[49] MDA, Government Relations Records. Series 3: Committees. Basil O'Connor to Sen. Charles Percy. September 3, 1968.

evaluation centers in hospitals throughout the nation in the 1960s. How-ever, in its annual report of 1972 titled "Maternal and Infant Health," the March of Dimes announced explicitly that it would broaden this program on birth defects prevention to include "our support of nationwide efforts to organize effective, regionalized intensive care for the high-risk pregnant woman, her fetus, and her critically ill newborn, or, in short, perinatal care." The "architect in the fight against polio" had inched his way toward a vision of healthy pregnancy, newborn intensive care, and prematurity prevention that has characterized the March of Dimes ever since. But just when Vir-ginia Apgar convened the Committee on Perinatal Health in 1972 that formulated the landmark *Toward Improving the Outcome of Pregnancy*, Basil O'Connor passed from the scene.

ENVOI

James Roosevelt once remarked that of the many historians who had por-trayed FDR's closest associates during his presidency, all of them were "notably silent" about Basil O'Connor. He remembered the two as both friends and partners, characterizing the March of Dimes as a living legacy of their friendship. Recalling the Depression-era song, "Brother, Can You Spare a Dime?" he claimed, "I always date the end of the depression by the call of The National Foundation to give a dime to the March of Dimes to rectify the inequalities between men created by disease."[50] Historians *have* been remarkably silent about Basil O'Connor, but that neglect is partly due to O'Connor's own closely-guarded habits. A recent biography of Louis Howe is titled *FDR's Shadow*, a characterization fully congruent with Howe's personal mission as political coach who seemed never to leave FDR's side. But if Howe was FDR's shadow, Basil O'Connor was his for-gotten man. Rexford Tugwell claimed that Howe regarded O'Connor with disdain, perhaps because in the centripetal movement of O'Connor's haughty independence he tended to loom as almost as large as Roosevelt's own good self. O'Connor's allegiance was indefatigable, for of all of Roos-evelt's associates, including the memoirists who looked back with affection and nostalgia over their years with FDR, Basil O'Connor was the single person (aside from Eleanor Roosevelt) who elevated the personal goals of FDR to further accomplishments after his death. O'Connor tenaciously stayed the course with Roosevelt, and he did so in Roosevelt's name: every publication and item of correspondence issued by the March of Dimes and

[50] MDA, Basil O'Connor Papers. Series 3: Dinner Programs. Seventy-fifth birthday tribute, 1966.

all of its chapters proudly bore the words "Franklin D. Roosevelt, Founder" until O'Connor's demise. The NFIP had eradicated a single disease, polio-myelitis; but in doing so, it "dramatized the idea of treating a person with a crippling disability so that he may be a self-sufficient member of society rather than an invalid who depends upon others."[51] Together, what Roos-evelt and O'Connor had accomplished was to *change the perception* of illness and disability by creating an environment for successful recovery and rein-tegration into society at Warm Springs and by empowering the effective coordination of financial, scientific, and medical resources for the general fight against disease and the promotion of public health through the March of Dimes.

Basil O'Connor died in Phoenix, Arizona on March 9, 1972 as he pre-pared to attend a March of Dimes grant meeting. Joseph Nee succeeded him as president, reorganized the March of Dimes board, and the Founda-tion began to embrace the perspective of perinatology for which O'Connor, Joseph Nee, and Charles Massey had set the groundwork in the 1960s. The obituaries that followed presented the usual encomia amid the casual swipes at O'Connor's idiosyncrasies. The *Montgomery Advertiser* called him a "mas-ter fund-raiser" but countered this with mention of his "excessive pride." The *Orlando Sentinel* equivocated by calling him a "gifted beggar." Proud of its distinguished alumnus, the *Dartmouth Alumni Magazine* described O'Connor as "a colorful figure, said by some to be the greatest fund-raiser ever." President Richard Nixon and former President Harry Truman sent condolence telegrams to Hazel O'Connor, and in a eulogy delivered at Rockefeller University on March 13, 1972 Jonas Salk said of O'Connor, "he became an institution."[52] This simple observation was the most fitting and insightful tribute of all, for despite the claims of "excessive pride," Basil O'Connor had sublimated his intense ambitions in an institution that had not only made America polio conscious but had eliminated the disease out-right. In a sense, the NFIP under O'Connor had indeed become a "flexible force," but not quite as he had envisioned. In any case, it had altered the way Americans approached disease, advancing fields of endeavor that ranged from virology and physical therapy to philanthropy, civil rights, and disabil-ity rights. Basil O'Connor was a self-made man, "only one remove from servitude" as he was wont to say, and when he first set out with Franklin

---

[51] MDA, Basil O'Connor Papers. Series 8: Biographical Data. Biographical profile, 1959.
[52] Ibid. Eulogies. Joseph Nee memo, March 14, 1972. Basil O'Connor, 1892–1972 Four Eulogies. *Montgomery Advertiser*, March 13, 1972. *Orlando Sentinel*, March 11, 1972. *Dartmouth Alumni Magazine*, May 1972, p. 81.

Roosevelt to defeat polio, all he had going for him, in the words of sports journalist Bob Considine, was "a yellow legal pad, a telephone, and a superb knowledge of the business and financial community."[53] But his potential unfolded in capabilities and circumstances richer than Considine's sound-bite might allow, and his strategic friendship with FDR was the central factor in his life. As with FDR, some found in O'Connor an admixture of "saint and scoundrel." But a scoundrel is a "mean, immoral, or wicked person," and O'Connor was never that. Jonas Salk, perhaps closer than anyone to O'Connor in the struggles of his later years, reflected that Basil O'Connor "was different from anybody else in the crowd. … He had it within his power to cause almost anything to happen."[54]

As for FDR, his presidential legacy is assured. One of the most revered figures in American politics, he has been reviled as a "traitor to his class" and idolized as "the savior of Western civilization." For neither Franklin nor Eleanor Roosevelt will there ever be a definitive summing up. Without question, FDR was the most famous "polio" of modern time. In his work at Warm Springs, he was described as "a builder" and called "prophetic." The impact of polio and life-changing disability on his political career will continue to command our curiosity and study. Eleanor Roosevelt was once asked, "Do you think your husband's illness has affected your husband's mentality?" She replied, "I am glad that question was asked. The answer is Yes. Anyone who has gone through great suffering is bound to have a greater sympathy and understanding of the problems of mankind."[55] This generalization now seems obvious but insufficient. After his personal letters were published in 1949, Mrs. Roosevelt felt free to reflect a bit more on her husband's suppression of his own emotions in the aftermath of polio: "As he came gradually to realize that he was not going to get any better, he faced great bitterness, I am sure, though he never mentioned it. The only thing that stands out in my mind as evidence of how he suffered when he finally knew that he would never walk again, was the fact that I never heard him mention golf from the day he was taken ill until the end of his life. That game epitomized to him the ability to be out of doors and to enjoy the use of his body. Though he learned to bear it, I am afraid it was always a tragedy."[56]

[53] MDA, Basil O'Connor Papers, Series 7: Memoranda. December 19, 1968.
[54] MDA, Oral History Records. Jonas Salk interview. September 8, 1984.
[55] MDA, Basil O'Connor Papers, Series 7, ibid.
[56] Roosevelt, Elliott, editor. *F.D.R. His Personal Letters.* Volume II, 1905–1928. New York: Duell, Sloan & Pearce, 1950. p. xviii.

Labor Secretary Frances Perkins felt that polio had sobered FDR, hardening his personal and political resolve, while Rexford Tugwell recognized his singular resolution as so completely fundamental to his character that not even paralysis could alter or extinguish it. He said, "I concluded that there had been no sudden conversion [after polio] and no change of character."[57] In evaluating Roosevelt, the personal, the public, and the political are so inextricably intertwined that it now seems almost inconceivable to disassociate the paralysis of economic depression from the paralysis of polio in the context of his biography. The self-created persona of "Doc Roosevelt" was refracted in the towering influence of his exemplary image as the four-term president who had seen the nation through depression and war. Basil O'Connor conflated these interpretations of character and magnified them all the more in consecrating the Roosevelt mythos in the service of the NFIP. Yet remembrance of the twin facts of Roosevelt's disability and his world-changing accomplishments did not require a Basil O'Connor to articulate them, for these are a common heritage. As Mario Cuomo stated of FDR in his keynote address to the 1984 Democratic National Convention, "He lifted himself from his wheelchair to lift this nation from its knees."[58]

More than a half century earlier, FDR himself had delivered a ringing speech at the Democratic National Convention. On June 28, 1928 in Houston, Texas FDR spoke before the assembled delegates to nominate Al Smith for President, the second such speech for Smith (the first, the "Happy Warrior" speech, occurred in 1924) that signaled quite clearly that Roosevelt was *not* defeated by polio and was still very much in the political game. The *New York Times* reported it as "A High-Bred Speech," lauding FDR as "a gentleman speaking to gentlemen" in characterizing Smith as a fair-minded and cultivated leader. Endorsing the intellectual and moral fitness of Smith for the presidency, FDR's language was remarkable for its soaring eloquence in describing the human qualities necessary for the office of President. Roosevelt orated:

> It is that quality of soul which makes a man loved by little children, by dumb animals; that quality of soul which makes him a strong help to all those in sorrow or in trouble; that quality which makes him not merely admired but loved by all the people—the quality of sympathetic understanding of the human heart, of real interest in one's fellowmen. Instinctively he senses the popular need because he himself has lived through the hardship, the labor and the sacrifice which must be endured by every man of heroic mold who struggles up to eminence from obscurity and low estate.

[57] Tugwell, Rexford G. *The Brains Trust.* New York: Viking, 1968. p. 64.
[58] Quoted in Smith, Jean Edward. *FDR.* New York: Random House, 2007.

Between him and the people is that subtle bond which makes him their champion and makes them enthusiastically trust him with their loyalty and their love.[59]

Save for mentioning the "low estate" of Al Smith's origins, Roosevelt might have been describing his own "quality of soul." It was something he instinctively understood.

In 2003, the *Journal of Medical Biography* published an article "What was the cause of Franklin Delano Roosevelt's paralytic illness?" entertaining the possibility that FDR suffered from, not poliomyelitis, but Guillain–Barré syndrome.[60,61] This article and a spate of books that have speculated on FDR's illnesses reveal the endless fascination with the medical condition of presidents, and in particular *this* President who had managed to disguise his disability in public life so to eventually defeat the disease that caused it. Despite the tantalizing speculation about FDR and Guillain–Barré syndrome, it remains impossible to make a corrective diagnosis retrospectively with any meaningful certainty absent a living patient. That Roosevelt *believed* his paralysis was due to polio, that he founded the organization that ultimately eradicated the disease, and that he remains one of the emblematic figures in the history of polio are alone sufficient to consign competing probabilities to a footnote rather than giving them undue prominence. In this regard it is noteworthy that the NFIP did not fail to consider Guillain–Barré syndrome in its program on polio patient aid. The NFIP's position on patients with Guillain–Barré syndrome who applied for aid was to allow its chapters to decide individually to award financial support to such patients based upon local factors rather than to develop a global policy. Basil O'Connor's medical directors judged this position as prudent and fair, especially since applicants with Guillain–Barré syndrome were very infrequent and thus not a burden to the Foundation.

After the O'Connor era the March of Dimes continued to memorialize FDR in celebratory events from time to time, as it did in 1982 with its "FDR Centennial Ball," a black-tie affair to re-create a 1930s birthday ball in a centennial observance of FDR's birth. James Roosevelt, then a board member of the March of Dimes, was chairman of the event. He remained a "goodwill ambassador" for the Foundation for many years, having

[59] Roosevelt, Elliott, editor. *F.D.R. His Personal Letters.* Volume II, 1905–1928. New York: Duell, Sloan & Pearce, 1950. p. 639.

[60] Goldman, Armond S. et al. 2003. "What was the cause of Franklin Delano Roosevelt's paralytic illness?" *Journal of Medical Biography.* 11: 232–240.

[61] Goldman, Armond S. et al. "Franklin Delano Roosevelt's (1882-1945) 1921 neurological disease revisited; the most likely diagnosis remains Guillain-Barré syndrome." *Journal of Medical Biography.* pii:0967772015605738, October 27, 2015 [Epub ahead of print].

maintained close ties with Basil O'Connor.[62] The Franklin Delano Roosevelt Memorial dedicated in Washington, DC in 1997 became a temporary battleground for disability rights activists who wanted FDR's disability unmasked for all time. The controversy over how best to represent his disability—and the very fact that it should be represented at all—has helped to hone our understanding about the political history of disability. By contrast with the monumental architecture of the Roosevelt Memorial in Washington and the Franklin D. Roosevelt Four Freedoms Park on Roosevelt Island in New York City, the Basil O'Connor Nature Trail at the Warm Springs Institute for Rehabilitation is diminutive, obscure, and barely noticeable. The March of Dimes, however, has honored his memory since the year of his death with the Basil O'Connor Starter Scholar Research Award. This grant award program was designed specifically to support promising research of young scientific investigators who are typically less able to obtain funding for innovative research at the beginning of their careers. Over the years, grants in this program have covered research in fields including pediatrics, obstetrics, neurobiology, biochemistry, pharmacology, biology, and genetics.

Charles Massey, who worked directly with O'Connor for over two decades, characterized the Roosevelt/O'Connor relationship in this fashion:

> Basil O'Connor was a pragmatic visionary. When FDR asked him to set up a foundation he had no qualms about using FDR's influence to get it done. I've always wondered how much of the credit for his success belonged to him and how much was attributable to the power of "that man in the White House." From the very outset the Foundation enjoyed tremendous support from Hollywood and the entertainment industry. Was that because of O'Connor's brilliance, or, was it, as some have suggested, because Hollywood catered to the President of the United States for its own reasons? Some cynics have speculated that the movie industry was simply concerned about anti-trust laws at that time and anxious to protect its turf. Whether that was the case I don't know, but I do know that Hollywood played a very big role in the Foundation's early success. As another case in point, the Foundation's first board of trustees was comprised of prestigious members drawn from the very top business and banking circles in the nation. They were willing to serve for only one reason. They were asked by FDR. In those days, at the bottom of every piece of the Foundation's literature were the words, "Franklin D. Roosevelt, Founder." I doubt that anyone knows whether the emphasis on FDR came from O'Connor's fertile mind or from the public relations specialists he hired. In truth it is difficult to fathom the relationship that existed between O'Connor and Roosevelt.[63]

[62] MDA, Public Relations Records. FDR Centennial Ball. FDRL, James Roosevelt Papers. Basil O'Connor correspondence.
[63] Massey, Charles. Personal interview. July 22, 2004.

Whether we can adequately plumb the imponderables of this fascinating partnership is left to our understanding of the documentary record and the contemporary momentum of the institutions that Roosevelt and O'Connor themselves created. Behind this realization lay the spirit of Warm Springs and the effervescence of the March of Dimes, and the two men—friends and partners—who were most responsible for formulating their vision and realizing their success.

APPENDIX

Letter from Franklin Delano Roosevelt to Basil O'Connor; November 10, 1942. Re: Authorization to use FDR's birthday for March of Dimes campaign, 1943

TRANSCRIPTION

The White House
Washington.   November 10, 1942

Dear Basil:

I have considered carefully your letter of November 5, 1942 requesting my views as to whether or not my birthday in January, 1943, should be publicly celebrated as heretofore in the fight against infantile paralysis.

In times such as these, that question cannot be decided entirely by itself. It must be judged in relation to other activities that we know are necessary to accomplish the one thing we all seek—Victory in this war.

At any time disease is a powerful enemy of man. In time of war, disease—particularly epidemic disease—is a factor which continuously gives us great concern. We know from history what can happen. We are constantly on the alert to prevent the start or spread of any of the epidemic diseases. There are no limits to which we will not go to accomplish that result. Such a policy is not only sound military strategy, but eminently humane.

And until it is definitely known how to prevent a disease from occurring or how to prevent it from spreading, the threat of that disease—if it is epidemic—is one of our greatest dangers, even though the actual number of cases at any given time may be relatively small. As long as there are some cases the danger exists.

The intensive fight we have been carrying on against infantile paralysis over these ten years—a short time in the history of any disease—has shown remarkable results, as you say; but more than that, it has followed a course which indicates that we will succeed in this struggle. Through intelligent planning and wise coordination, we are prepared, if necessary, for a long time fight against this disease.

I feel as you do, that any interruption in this work would be extremely inadvisable unless absolutely necessary. More than that, I also think that such a fight as that being waged [page 2:] against infantile paralysis or any other as yet uncontrolled disease is an essential part of the main struggle in which we all are engaged—a struggle to make tomorrow's world a better world in which to live. While we fight this global war, we must see to it that the health of our children is preserved and protected so that they may enjoy that better world—for tomorrow's America will be as strong as today's children. We must help them win their Victory over disease today.

As I have said in the past and repeat now—nothing is closer to my heart than the health of our boys and girls and young men and young women. To me it is one of the front lines of our National Defense.

I feel strongly, therefore, that the work of the National Foundation must be continued and I am happy to have it use my birthday in its 1943 fund-raising drive.

With my best wishes for a successful campaign, I am

Sincerely yours,
[signed:] Franklin Roosevelt

Mr. Basil O'Connor,
President, The National Foundation for Infantile Paralysis, Inc.,
120 Broadway
New York City.

BIBLIOGRAPHY

ARCHIVES AND SPECIAL COLLECTIONS

Franklin D. Roosevelt Presidential Library; Hyde Park, New York
    Anna Eleanor Roosevelt Papers
    Basil O'Connor Papers
    Basil O'Connor Collection
    Francis P. Corrigan Papers
    Franklin D. Roosevelt Library, Inc. Records
    Franklin D. Roosevelt Memorial Foundation Records
    Franklin D. Roosevelt: Papers as President, President's Secretary's File
    Franklin D. Roosevelt: President's Personal File
    Franklin D. Roosevelt: Papers pertaining to Family, Business and Personal
    James Roosevelt Papers
    O'Connor and Farber Records
    Rexford G. Tugwell Papers
    Robert D. Graff Papers
    Social Entertainments, Office of the Chief of Records
    Stephen T. Early Papers
    Vertical File: Basil O'Connor
    William D. Hassett Papers
March of Dimes Archives; White Plains, New York
    Basil O'Connor Papers
    Chapter Administration Records
    Conferences and Meetings Records
    Fund Raising Records
    Georgia Warm Springs Foundation Records
    Government Relations (Foreign) Records
    Grant Records (CRBS)
    History of the National Foundation for Infantile Paralysis Records
    Media and Publications Records
    Medical Program Records
    Oral History Records
    Photography Collection Public Relations Records
    Radio Spot Announcements Records
    Surveys and Studies Records
Rauner Special Collections Library, Dartmouth College; Hanover, New Hampshire
    The Papers of Basil O'Connor, Collection #MS-5Q2
American Red Cross, Hazel Braugh Records Center and Archives; Lorton, Virginia
    Historical Biographies
New York State Library, Manuscripts and Special Collections; Albany, New York
    Basil O'Connor Papers
Roosevelt Warm Springs Institute for Rehabilitation Archives

PUBLISHED WORKS

Altenbaugh, R.J., 2006. Where are the disabled in the history of education? The impact of polio on sites of learning. History of Education 35 (6), 705–730.
Altenbaugh, R., 2015. The Last Children's Plague: Poliomyelitis, Disability, and Twentieth Century American Culture. Palgrave Macmillan.

Altman, L.K., 1987. Who Goes First?: The Story of Self-Experimentation in Medicine. Random House, New York.

Apgar, V., 1968. Birth defects: their significance as a public health problem. JAMA: The Journal of the American Medical Association 204 (5), 371–374.

Baghdady, G., Maddock, J.M., 2008. Marching to a different mission. Stanford Social Innovation Review 59–65.

Baker, J.P., 1996. The Machine in the Nursery: Incubator Technology and the Origins of Newborn Intensive Care. Johns Hopkins UP, Baltimore.

Baltzell, E.D., 1964. The Protestant Establishment: Aristocracy & Caste in America. Vintage, New York.

Benison, S., Rivers, T.M., 1967. Tom Rivers: Reflections on a Life in Medicine and Science. M.I.T., Cambridge.

Benison, S., 1967. The enigma of poliomyelitis: 1910. In: Hyman, H.M., Levy, L.W. (Eds.), Freedom and Reform: Essays in Honor of Henry Steele Commager. Harper & Row, New York.

Benison, S., 1972. The history of polio research in the United States: appraisal and lessons. In: Holton, G.J. (Ed.), The Twentieth Century Sciences: Studies in the Biography of Ideas. Norton, New York.

Benison, S., 1982. International medical cooperation: Dr. Albert Sabin, live polio virus, and the Soviets. Bulletin of the History of Medicine 56 (4), 460–483.

Benison, S., 1974. Poliomyelitis and the Rockefeller Institute: social effects and institutional response. Journal of the History of Medicine 29, 74–93.

Benison, S., 1965. Reflections on oral history. The American Archivist 28 (1), 71–77.

Berg, R.H., 1946. The Challenge of Polio. Dial, New York.

Berg, R.H., 1948. Polio and Its Problems. J.B. Lippincott, Philadelphia.

Berlin, I., 2014. Personal Impressions. 1980. Princeton UP, Princeton.

Bernstein, I., 1960. The Lean Years: A History of the American Worker, 1920–1933. Penguin, Baltimore.

Black, C., 2003. Franklin Delano Roosevelt: Champion of Freedom. Public Affairs, New York.

Black, K., 1996. In the Shadow of Polio: A Personal and Social History. Addison-Wesley Pub., New York.

Blakeslee, A.L., 1956. Polio and the Salk Vaccine: What You Should Know About It. Grosset & Dunlap, New York.

Bishop, J., 1974. FDR's Last Year: April 1944–April 1945. Morrow, New York.

Bourgeois, S., 2013. Genesis of the Salk Institute: The Epic of Its Founders. University of California.

Boyd, T.E., 1953. Immunization against poliomyelitis. Bacteriological Reviews 17, 339–448.

Boyer, P.S., 1985. By the Bomb's Early Light: American Thought and Culture at the Dawn of the Atomic Age. Pantheon, New York.

Brand, H.W., 2008. Traitor to His Class: The Privileged Life and Radical Presidency of Franklin Delano Roosevelt. Doubleday, New York.

Brandt, A.M., 1978. Polio, politics, publicity, and duplicity: ethical aspects in the development of the Salk vaccine. International Journal of Health Services 8 (2), 257–270.

Brock, P., 2008. Charlatan: America's Most Dangerous Huckster, the Man Who Pursued Him, and the Age of Flimflam. Crown, New York.

Bruenn, H.G., 1970. Clinical notes on the illness and death of President Franklin D. Roosevelt. Annals of Internal Medicine 72 (4), 579–591.

Buhite, R.D., Levy, D.W. (Eds.), 1993. FDR's Fireside Chats. Penguin, Harmondsworth.

Buhle, P., Jones, S., 2010. FDR and the New Deal for Beginners. For Beginners, Danbury, CT.

Burns, K., 2014. The Roosevelts: An Intimate History. Florentine Films. DVD.

Buirski, N., 2013. Afternoon on a Faun: Tanaquil Le Clercq. Augusta Films. DVD.

Burke Jr., D.M., Burke, O.A., 2005. Warm Springs. Arcadia, Charleston.

Burns, J.M., 1956. Roosevelt: The Lion and the Fox. Harcourt, Brace & World, New York.

Burns, J.M., 1970. Roosevelt: The Soldier of Fortune, 1940–1945. Harcourt, New York.

Burns, K., 2015. Cancer: The Emperor of All Maladies. Florentine Films. DVD.

Camus, A., 2006. Camus at Combat: writing 1944–1947. In: Lévi-Valensi, J. (Ed.), Trans. Arthur Goldhammer. Princeton UP, Princeton.

Carter, K.F., 2001. Trumpets of attack: collaborative efforts between nursing and philanthropies to care for the child crippled with polio, 1930 to 1959. Public Health Nursing 18 (4), 253–261.

Carter, R., 1966. Breakthrough: The Saga of Jonas Salk. Trident, New York.

Carter, R., 1961. The Gentle Legions. Doubleday, Garden City.

Caute, D., 1978. The Great Fear: The Anti-Communist Purge Under Truman and Eisenhower. Simon and Schuster, New York.

Cellular Biology Nucleic Acids and Viruses, 1957. Special Publication of The New York Academy of Sciences, vol. 5.

Cerf, B., 1977. At Random: The Reminiscences of Bennett Cerf. Random House, New York.

Chapell, E.P., Hume, J.F., 2008. A Black Oasis: Tuskegee Institute's Fight Against Infantile Paralysis, 1941–1975.

Chappell, E., 1960. On the Shoulders of Giants: The Bea Wright Story. Chilton, Philadelphia.

Codr, D., 2014. Arresting Monstrosity: Polio, Frankenstein, and the Horror Film. Publications of the Modern Language Association 129 (2), 171–187.

Cohn, V., 1955. Four Billion Dimes. Minneapolis Star and Tribune, Minneapolis.

Cohn, V., 1975. Sister Kenny: The Woman Who Challenged the Doctors. U of Minnesota, Minneapolis.

Collier, P., Horowitz, D., 1994. The Roosevelts: An American Saga. Simon & Schuster, New York.

Colt, S., 2009. The Polio Crusade. PBS American Experience. DVD.

Cook, B.W., 1992. Eleanor Roosevelt, Volume 1: 1884–1933, vol. 1. Penguin, New York.

Cook, B.W., 1999. Eleanor Roosevelt, Volume 2: The Defining Years, 1933–1938, vol. 2. Penguin, New York.

Costigliola, F., 2012. Roosevelt's Lost Alliances: How Personal Politics Helped Start the Cold War. Princeton UP, Princeton.

Creager, A.N.H., 2002. The Life of a Virus: Tobacco Mosaic Virus as an Experimental Model, 1930–1965. University of Chicago, Chicago.

Creager, A.N.H., 2008. Mobilizing biomedicine: virus research between Lay Health Organizations and the United States Federal Government, 1935–1955. In: Hannaway, C. (Ed.), Biomedicine in the 20th Century: Practices, Policies, and Politics. IOS, Amsterdam, pp. 171–201.

Cutlip, S.M., 1990. Fund Raising in the United States, Its Role in America's Philanthropy. Transaction, New Brunswick.

Daniel, T.M., Robbins, F.C., 1997. Polio. University of Rochester, Rochester.

De Kruif, P., 1939. Activities of the NFIP in the Field of Virus Research. NFIP.

De Kruif, P., 1938. The Fight for Life. Harcourt Brace, New York.

De Kruif, P., 1949. Life among the Doctors. Harcourt Brace, New York.

De Kruif, P., 1926. Microbe Hunters. Harcourt Brace, New York.

De Kruif, P., 1922. Our Medicine Men. Century, New York.

De Kruif, P., 1962. The Sweeping Wind: A Memoir. Harcourt Brace, London.

Duffy, J., 1987. Franklin Roosevelt: ambiguous symbol for disabled Americans. Midwest Quarterly 113–135.

Dulles, F.R., 1950. The American Red Cross: A History. Harper & Brothers, New York.

Engel, L., Aug. 1955. The Salk vaccine: what caused the mess? Harper's Magazine 4, 1–32.

Erenberg, L.A., 1999. Swingin' the Dream: Big Band Jazz and the Rebirth of American Culture. University of Chicago Press, Chicago.

Excerpta Medica, comp., 1970. Congenital Malformations: 3rd International Conference on Congenital Malformations. J.B. Lippincott, Amsterdam.

Fairchild, A.L., 2001. The polio narratives: dialogues with FDR. Bulletin of the History of Medicine 75 (3), 488–534.

Farley, J.A., 1938. Behind the Ballots: The Personal History of a Politician. Harcourt Brace, New York.

Farley, J.A., 1948. Jim Farley's Story: The Roosevelt Years. Whittlesey House, New York.

Fenster, J.M., 2009. FDR's Shadow: Louis Howe, the Force That Shaped Franklin and Eleanor Roosevelt. Palgrave Macmillan, New York.

Finkelstein, L. (Ed.), 1953. Thirteen Americans: Their Spiritual Autobiographies. The Institute for Religious and Social Studies, New York.

Foertsch, J., 2008. Bracing Accounts: The Literature and Culture of Polio in Postwar America. Fairleigh Dickinson UP, Madison.

Francis Jr., T., 1957. Evaluation of the 1954 Field Trial of Poliomyelitis Vaccine Final Report. Edwards Brothers, Ann Arbor.

Freidel, F., 1954. Franklin D. Roosevelt: The Ordeal. Little Brown, Boston.

Friedenberg, Z.B., 2009. Franklin D. Roosevelt: his poliomyelitis and orthopaedics. The Journal of Bone and Joint Surgery 91 (7), 1806–1813.

Gabler, N., 1988. An Empire of Their Own: How the Jews Invented Hollywood. Doubleday, New York.

Gardner, D.S., 2009. Roosevelt House at Hunter College: The Story of Franklin and Eleanor's New York City Home. Gilder Lehrman Institute of American History, New York.

Geddes, D.P. (Ed.), 1945. Franklin Delano Roosevelt: A Memorial. Pitman, New York.

Glendon, M.A., 2001. A World Made New: Eleanor Roosevelt and the Universal Declaration of Human Rights. Random House, New York.

Goldberg, R.T., 1981. The Making of FDR: Triumph over Disability. Abt, Cambridge.

Goldman, A.S., et al., 2003. What was the cause of Franklin Delano Roosevelt's paralytic illness? Journal of Medical Biography 11, 232–240.

Goldman, A.S., et al., October 27, 2015. Franklin Delano Roosevelt's (1882–1945) 1921 neurological disease revisited; the most likely diagnosis remains Guillain-Barré syndrome. Journal of Medical Biography. pii:0967772015605738 [Epub ahead of print].

Goodwin, D.K., 1995. No Ordinary Time: Franklin and Eleanor Roosevelt: The Home Front in World War II. Simon & Schuster, New York.

Gould, J., Hickok, L., 1972. Walter Reuther: Labor's Rugged Individualist. Dodd, Mead, New York.

Gould, J., 1960. A Good Fight, The Story of FDR's Conquest of Polio. Dodd, Mead, New York.

Gould, T., 1995. A Summer Plague: Polio and Its Survivors. Yale UP, New Haven.

Greidanus, T., 2010. The Shot Felt 'Round the World. Steeltown Entertainment Project. DVD.

Gudakunst, D.W., 1942. The Importance of Research. National Foundation for Infantile Paralysis, New York.

Gunn, S.M., Platt, P.S., 1945. Voluntary Health Agencies, an Interpretive Study. Ronald, New York.

Hansen, B., 2009. Picturing Medical Progress from Pasteur to Polio: A History of Mass Media Images and Popular Attitudes in America. Rutgers UP, New Brunswick.

Harris, M.J., 1984. The Homefront: America During World War II. Putnam, New York.

Hassett, W.D., 1960. Off the Record with F.D.R., 1942–1945. Allen & Unwin, London.

Hawke, C., 2015. Influenza and Influence: Behind the Scenes with Salk and Sabin in 1976. (Unpublished).

Hawkins, L.C., Lomask, M., 1956. The Man in the Iron Lung: The Frederick B. Snite, Jr., Story. Doubleday, Garden City.

Heaton, A., July 1953. A Friend–and Partner. Good Housekeeping, p. 209.

Hogan, A.J., 2013. Locating genetic disease: the impact of clinical nosology on biomedical conceptions of the human genome (1966–1990). New Genetics and Society 32 (1), 78–96.

Hogan, A.J., 2014. The 'morbid anatomy' of the human genome: tracing the observational and representational approaches of postwar genetics and biomedicine. Medical History 58 (03), 315–336.

Horstmann, D.M., 1985. The poliomyelitis story: a scientific hegira. Yale Journal of Biology and Medicine 58, 79–90.

Horstmann, D.M., 1991. The Sabin live poliovirus vaccination trials in the USSR. Yale Journal of Biology and Medicine 501–502.

Houck, D.W., 2002. FDR and Fear Itself: The First Inaugural Address. Texas A & M UP, College Station.

Hurd, C., 1965. When the New Deal Was Young and Gay: FDR & His Circle. Hawthorn.

International Medical Congress, comp., 1963. Congenital Defects: Papers and Discussions Presented at the First Inter-American Conference. J.B. Lippincott, New York.

International Medical Congress, comp., 1961. Congenital Malformations: Papers and Discussions Presented at the First International Conference. J.B. Lippincott, New York.

International Medical Congress, comp., 1964. Congenital Malformations: Papers and Discussions Presented at the Second International Conference. J.B. Lippincott, New York.

Jacobs, C., 2015. Jonas Salk: A Life. Oxford UP, New York.

Josephson, E., November 1956. Why the dimes march on. Nation 10, 361–364.

Judson, H.F., 1979. The Eighth Day of Creation: Makers of the Revolution in Biology. Simon and Schuster, New York.

Katznelson, I., 2013. Fear Itself: The New Deal and the Origins of Our Time. Liveright, New York.

Kennedy, D.M., 1999. Freedom from Fear: The American People in Depression and War, 1929–1945. Oxford UP, New York.

Kenny, E., Ostenso, M., 1951. And They Shall Walk: The Life Story of Sister Elizabeth Kenny. Hale, London.

Kenny, E., 1955. My Battle and Victory: History of the Discovery of Poliomyelitis as a Systemic Disease. Hale, London.

Kershaw, I., 2007. Fateful Choices: Ten Decisions That Changed the World, 1940–1941. Penguin, New York.

Klein, A.E., 1972. Trial by Fury: The Polio Vaccine Controversy. Scribner, New York.

Klingaman, W.K., 1988. 1941: Our Lives in a World on the Edge. Harper & Row, New York.

Kluger, J., 2005. Splendid Solution: Jonas Salk and the Conquest of Polio. Penguin, New York.

Koppes, C.R., Black, G.D., 1987. Hollywood Goes to War: How Politics, Profits, and Propaganda Shaped World War II Movies. Berkeley & Los Angeles: University of California.

Lambert, S.M., Markel, H., 2000. Making history: Thomas Francis, Jr., MD, and the Salk poliomyelitis vaccine field trial. Archives of Pediatrics & Adolescent Medicine 154 (5), 512–517.

Lasch, C., 1979. The Culture of Narcissism: American Life in an Age of Diminishing Expectations. Norton, New York.

Lash, J.P., 1971. Eleanor and Franklin: The Story of Their Relationship, Based on Eleanor Roosevelt's Private Papers. Norton, New York.

Lash, J.P., 1972. Eleanor: The Years Alone. Norton, New York.

Leavitt, R.K., 1949. Common Sense About Fund Raising. Stratford, New York.

Leland, W.G., 1955. The creation of the Franklin D. Roosevelt library: a personal narrative. The American Archivist 18 (1), 11–29.

Leuchtenburg, W.E., 1985. In the Shadow of FDR: From Harry Truman to Ronald Reagan. Cornell UP, Ithaca.

Levine, L.W., Levine, C.R., 2002. The People and the President: America's Conversation with FDR. Beacon, Boston.

Lingeman, R.R., 1970. Don't You Know There's a War On?: The American Home Front, 1941–1945. Perigee, New York.

Lippman Jr., T., 1978. The Squire of Warm Springs: FDR in Georgia, 1924–1945. Playboy Paperbacks, Chicago.

Littlefield, J., Grouchy, J.D., 1978. Birth Defects. Excerpta Medica, Amsterdam.

Liv Estrup, 2010. Flying without Wings: Life with Arnold Beisser. Liv Estrup. DVD.

Live Poliovirus Vaccines, 1960. Papers Presented and Discussions Held at the Second International Conference on Live Poliovirus Vaccines. Pan American Health Organization, Washington, DC.

Loewen, J.W., 2007. Lies My Teacher Told Me: Everything Your American History Textbook Got Wrong, revised ed. Simon & Schuster, New York.

Logan, L.H., 2005. A Summer Without Children: An Oral History of Wythe County, Virginia's 1950 Polio Epidemic. In: Stevan, R. (Ed.), Wytheville: Town of Wytheville Dept. of Museums, Jackson.

Lomazow, S., Fettmann, E., 2009. FDR's Deadly Secret. Public Affairs, New York.

Mawdsley, S.E., 2013. Balancing risks: childhood inoculations and America's response to the provocation of paralytic polio. Social History of Medicine 26 (4), 759–778.

Mawdsley, S.E., 2010. "Dancing on eggs": Charles H. Bynum, racial politics, and the National Foundation for Infantile Paralysis, 1938–1954. Bulletin of the History of Medicine 84 (2), 217–247.

Mawdsley, S.E., 2016. Selling Science: Polio and the Promise of Gamma Globulin. Rutgers UP, New Brunswick.

Meacham, J., 2003. Franklin and Winston: An Intimate Portrait of an Epic Friendship. Random House, New York.

Minchew, K., 2016. A President in Our Midst: Franklin Delano Roosevelt in Georgia. University of Georgia, Athens.

Mitman, G., 2010. "The Color of Money: Campaigning for Health in Black and White America." Imagining Illness: Public Health and Visual Culture. In: Serlin, D. (Ed.), University of Minnesota, Minneapolis, pp. 40–61.

Moore, M., 1967. The Complete Poems of Marianne Moore. Macmillan, New York.

National Foundation for Infantile Paralysis, 1941a. Infantile Paralysis: A Symposium Delivered at Vanderbilt University. NFIP.

National Foundation for Infantile Paralysis, 1941b. The Story of the National Foundation for Infantile Paralysis. NFIP, New York.

Nekola, A., Kirkpatrick, B., 2010. Cultural policy in American music history: Sammy Davis, Jr., vs. juvenile delinquency. Journal of the Society for American Music 4 (1), 33–58.

Nevins, A., 1961. The Gateway to History, revised ed. Doubleday, Garden City.

O'Farrell, B., 2010. She Was One of Us: Eleanor Roosevelt and the American Worker. Cornell UP, Ithaca.

Offit, P.A., 2005. The Cutter Incident: How America's First Polio Vaccine Led to the Growing Vaccine Crisis. Yale UP, New Haven.

Offit, P.A., 2007. Vaccinated: One Man's Quest to Defeat the World's Deadliest Diseases. Smithsonian, Washington, DC.

Oppewal, S., 1997. Sister Kenny, an Australian nurse, and treatment of poliomyelitis victims. Image: The Journal of Nursing Scholarship 83–87.

Oshinsky, D.M., 2005. Polio: An American Story. Oxford UP, Oxford.

Paul, J.R., 1971. A History of Poliomyelitis. Yale UP, New Haven.

Pederson, T., 2013. Turning on a dime: the 75th anniversary of America's march against polio. The FASEB Journal 27 (7), 2533–2535.

Perkins, F., 1946. The Roosevelt I Knew. Viking, New York.

Perrett, G., 1973. Days of Sadness, Years of Triumph; the American People, 1939–1945. Penguin, Baltimore.

Persico, J.E., 2001. Roosevelt's Secret War: FDR and World War II Espionage. Random House, New York.

Phillips, C.B.H., 1969. From the Crash to the Blitz, 1929–1939. New York Times, New York.

Phillips, H.B. (Ed.), 1962. Felix Frankfurter Reminisces: An Intimate Portrait as Recorded in Talks with Dr. Harlan B. Phillips. Doubleday, New York.

Plokhy, S.M., 2010. Yalta: The Price of Peace. Viking, New York.

Poliomyelitis, 1961. Papers and Discussions Presented at the Fifth International Poliomyelitis Conference. J.B. Lippincott, Philadelphia.

Poliomyelitis, 1949. Papers and Discussions Presented at the First International Poliomyelitis Conference. J.B. Lippincott, Philadelphia.

Poliomyelitis, 1958. Papers and Discussions Presented at the Fourth International Poliomyelitis Conference. J.B. Lippincott, Philadelphia.

Poliomyelitis, 1952. Papers and Discussions Presented at the Second International Poliomyelitis Conference. J.B. Lippincott, Philadelphia.

Poliomyelitis, 1955. Papers and Discussions Presented at the Third International Poliomyelitis Conference. J.B. Lippincott, Philadelphia.

Quinn, S., 2008. Furious Improvisation: How the WPA and a Cast of Thousands Made High Art out of Desperate Times. Walker, New York.

Reagan, L.J., 2010. Dangerous Pregnancies: Mothers, Disabilities, and Abortion in Modern America. University of California, Berkeley.

Rhodes, J., 2013. The End of Plagues: The Global Battle Against Infectious Disease. Palgrave Macmillan, New York.

van Rijn, G., 1997. Roosevelt's Blues: African-American Blues and Gospel Songs on FDR. University of Mississippi, Jackson.

Rivers, T.M., 1928. Filterable Viruses. Williams & Wilkins, Baltimore.

Rivers, T.M. (Ed.), 1952. Viral and Rickettsial Infections of Man, second ed. J. B. Lippincott, Philadelphia.

Rogers, N., 1992. Dirt and Disease: Polio Before FDR. Rutgers UP, New Brunswick.

Rogers, N., 2009. Polio chronicles: warm springs and disability politics in the 1930s. Asclepio: Revista De Historia De La Medicina Y De La Ciencia 61 (1), 143–174.

Rogers, N., 2014. Polio Wars: Sister Kenny and the Golden Age of American Medicine. Oxford UP, New York.

Rogers, N., 2007. Race and the politics of polio: Warm Springs, Tuskegee, and the March of Dimes. American Journal of Public Health 97 (5), 784–795.

Rohmer, P.H., 1913. Epidemic Infantile Paralysis. William Wood, New York.

Roosevelt, E., 2001. In: Emblidge, D. (Ed.), My Day: The Best of Eleanor Roosevelt's Acclaimed Newspaper Columns, 1936–1962. Da Capo, New York.

Roosevelt, E., 1949. This I Remember. Harper & Brothers, New York.

Roosevelt, E., 2012. Tomorrow Is Now. 1963. Reprint. Penguin, New York.

Roosevelt, F.D., 2009. Looking Forward. 1933. Reprint. Simon & Schuster, New York.

Roosevelt, F.D., Hassett, W.D., 1937-1940. In: Rosenman, S.I. (Ed.), The Public Papers and Addresses of Franklin D. Roosevelt. Random House, New York.

Roosevelt, F.D., 1947. In: Roosevelt, E. (Ed.), F.D.R.: His Personal Letters: The Early Years, vol. 1. Duell, Sloan & Pearce.

Roosevelt, F.D., 1999. In: Grafton, J. (Ed.), Great Speeches. Dover Publications, Mineola.

Roosevelt, F.D., 1973. In: Venkataramani, M.S. (Ed.), The Sunny Side of FDR. Ohio UP, Athens.

Roosevelt, J., Shalett, S., 1959. Affectionately, FDR: A Son's Story of a Lonely Man. Harcourt Brace, New York.

Rose, D.W., 2003. March of Dimes. Arcadia, Charleston.

Rose, D.W., 2007. Robert A. Good, the March of Dimes, and immunodeficiency: an historical perspective. Immunological Research 38 (1–3), 51–54.

Rosenberg, C., 1963. Martin Arrowsmith, the scientist as hero. American Quarterly Fall 447–458.

Roth, P., 2010. Nemesis. Houghton Mifflin Harcourt, Boston.

Roth, P., 2004. The Plot Against America. Houghton Mifflin, New York.

Rothman, D.J., 1997. "The Iron Lung and Democratic Medicine." Beginnings Count: The Technological Imperative in American Health Care. Oxford UP, pp. 42–66.

Rowe, D.E., Schulmann, R.J., 2007. Einstein on Politics: His Private Thoughts and Public Stands on Nationalism, Zionism, War, Peace, and the Bomb. Princeton UP, Princeton.

Salk, J., 1972. Man Unfolding. Harper & Row, New York.

Schary, D., 1958. Sunrise at Campobello. New American Library, New York.

Schlesinger, A.M., 1958. The Coming of the New Deal: 1933–1935, vol. 2. Houghton Mifflin, Boston.

Schlesinger, A.M., 1985. The Crisis of the Old Order: 1919–1933, vol. 1. Houghton Mifflin, Boston.

Schlesinger, A.M., 1960. The Politics of Upheaval: 1935–1936, vol. 3. Houghton Mifflin, Boston.

Seavey, N.G., Jane, S.S., Wagner, P., 1998. A Paralyzing Fear: The Triumph Over Polio In America. TV, New York.

Shell, M., 2005. Polio and Its Aftermath: The Paralysis of Culture. Harvard UP, Cambridge.

Sherwood, R.E., 1948. Roosevelt and Hopkins: An Intimate History. Harper & Brothers, New York.

Shreve, S.R., 2007. Warm Springs: Traces of a Childhood at FDR's Polio Haven. Houghton Mifflin, Boston.

Sills, D.L., 1957. The Volunteers: Means and Ends in a National Organization. Free, Glencoe.

Sink, A.E., 1998. The Grit Behind the Miracle. University of America, Lanham.

Skinner, G., 2005. The Christmas House: How One Man's Dream Changed the Way We Celebrate Christmas. Novato: New World Library.

Skloot, R., 2010. The Immortal Life of Henrietta Lacks. Crown, New York.

Smith, I.T., 1949. "Dear Mr. President…": The Story of Fifty Years in the White House Mail Room. J. Messner, New York.

Smith, I.T., May 1954. O Pioneers!. New Yorker 8.

Smith, J.S., 1990. Patenting the Sun: Polio and the Salk Vaccine. William Morrow, New York.

Smith, J.E., 2007. FDR. Random House, New York.

Steeholm, C., Steeholm, H., 1950. The House at Hyde Park. Viking, New York.

Stowe, D.W., 1996. Swing Changes: Big Band Jazz in New Deal America. Harvard UP, Cambridge.

Tobin, J., 2013. The Man He Became: How FDR Defied Polio to Win the Presidency. Simon & Schuster, New York.

Tugwell, R.G., 1968. The Brains Trust. Viking, New York.

Tugwell, R.G., 1957. The Democratic Roosevelt: A Biography of Franklin D. Roosevelt. Penguin, Baltimore.

Tully, G., 1949. F.D.R.: My Boss. Scribner, New York.

Villas, J., 2002. Between Bites: Memoirs of a Hungry Hedonist. Wiley, New York.

Walker, T., 1951. Journey Together. McKay, New York.

Walker, T., 1950. Rise Up and Walk. Dutton, New York.

Walker, T., 1954. Roosevelt and the Fight Against Polio. Rider, London.

Walker, T., 1953. Roosevelt and the Warm Springs Story. Wyn, New York.

Ward, G.C., 1995. Closest Companion: The Unknown Story of the Intimate Friendship between Franklin Roosevelt and Margaret Suckley. Houghton Mifflin, Boston.

Ward, G.C., 1989. A First-Class Temperament: The Emergence of Franklin Roosevelt. Harper & Row, New York.

West, H.F., 1966. The Impecunious Amateur Looks Back: The Autobiography of a Bookman. Westholm Publications, Hanover.

Widukind, L., Motulsky, A.G., Ebling, F.J.G. (Eds.), 1974. Birth Defects: Proceedings of the Fourth International Conference. Excerpta Medica, Amsterdam.

Williams, G., 2013a. Paralyzed with Fear: The Story of Polio. Palgrave Macmillan, London.

Williams, M.B., 2013b. City of Ambition: FDR, La Guardia, and the Making of Modern New York. Norton, New York.

Wilson, D.J., 2009a. And they shall walk: ideal versus reality in polio rehabilitation in the United States. Asclepio: Revista De Historia De La Medicina Y De La Ciencia 61 (1), 175–192.

Wilson, D.J., 2014. Basil O'Connor, the National Foundation for Infantile Paralysis and the Reorganization of Polio Research in the United States, 1935–41. Journal of the History of Medicine and Allied Sciences 70 (3), 394–424.

Wilson, D.J., 2005. Living with Polio: The Epidemic and Its Survivors. U of Chicago, Chicago.

Wilson, D.J., 2009b. Polio: Biographies of Disease. Santa Barbara, Greenwood.

Wilson, D.J., 2008. Psychological trauma and its treatment in the polio epidemics. Bulletin of the History of Medicine 82 (4), 848–877.

Wilson, J.R., 1963. Margin of Safety: The Story of Poliomyelitis Vaccine. Doubleday, London.

Wohlers, C., 2013. The Shot That Saved the World. Smithsonian Network. DVD.

Wooten, H.G., 2009. The Polio Years in Texas: Battling a Terrifying Unknown. Texas A & M UP, College Station.

Young, C.J., 2011. Proclamations and the Founding Father Presidents, 1789–1825. Federal History Journal 3, 80–90 Society for History in the Federal Government. January 2011.

ACADEMIC THESES AND DISSERTATIONS

Matysiak, A., 2005. The Development of an Oral Vaccine Against Poliomyelitis (thesis). George Washington University.

Mawdsley, S.E., 2011. Fighting Polio: Selling the Gamma Globulin Field Trials, 1950–1953 (thesis). University of Cambridge.

Mawdsley, S.E., 2006. Harnessing the Power of People: The Fundraising Efforts of the National Foundation for Infantile Paralysis, 1938–1954 (thesis). University of Alberta.

Mawdsley, S.E., 2008. Polio and Prejudice: Charles Hudson Bynum and the Racial Politics of the National Foundation for Infantile Paralysis, 1938–1954 (thesis). University of Alberta.

Moeschen, S.C., 2005. Benevolent Actors and Charitable 'Objects': Physical Disability and the Theatricality of Charity in Nineteenth and Twentieth-Century America (thesis). Northwestern University.

Scheffler, R.W., 2014. Cancer Viruses and the Construction of Biomedicine in the United States from 1900 to 1980 (thesis). Yale University.

Smith, D.L., 2015. Sisters of Mercy: The Walking Nuns' Siouxland Journey and Experiences in Nursing (1890–1965) (dissertation). South Dakota State University.

Wooten, H.G., 2006. The Polio Years in Harris and Galveston Counties, 1930–1962 (thesis). University of Texas Medical School.

INDEX

'*Note*: Page numbers followed by "f" indicate figures.'

Printed and bound by CPI Group (UK) Ltd, Croydon, CR0 4YY

03/10/2024

01040421-0002